# The CIM Debacle

**Springer**
*Singapore*
*Berlin*
*Heidelberg*
*New York*
*Barcelona*
*Budapest*
*Hong Kong*
*London*
*Milan*
*Paris*
*Santa Clara*
*Tokyo*

# The CIM Debacle

Methodologies to Facilitate Software Interoperability

VALDEW SINGH

 Springer

**Dr. Valdew Singh**
CIM Centre
German-Singapore Institute
Nanyang Polytechnic
10 Science Centre Road
Singapore 609079

**ISBN 981-3083-21-2**

© Springer-Verlag Singapore Pte. Ltd. 1997
Printed in Singapore

The publisher makes no representation, express or implied, with regard to the accuracy of the information contained in this book and cannot accept any legal responsibility or liability for any errors or omissions that may be made.

Typesetting: Camera-ready by Author
5 4 3 2 1 0

# Dedication

*To my dear wife, Magdalene, who has been my constant source of inspiration, support and encouragement;*

and

*My late father, Banta Singh, who inculcated in me the belief in the pursuit of knowledge for self-fulfilment and self-realisation and to serve the betterment of others.*

# Dedication

# Foreword

It is now widely accepted that a new breed of systems modelling and engineering tools is required which can support the holistic conceptualisation of integrated manufacturing systems and their rapid and effective development on an incremental basis. However in practice this is extremely difficult to achieve. Invariably:

(i) manufacturing enterprises are extremely complex organisms and require interdisciplinary teams of people to support their life-cycle engineering;

(ii) as yet suitable formal analysis techniques do not exist which in a consistent way can describe, support and maintain links throughout the chain: *definition of market/environmental opportunity, risk and need; definition of the business processes in an enterprise; definition of organisation structures to be adopted by an enterprise; definition and selection of the social and technical systems deployed; specification, co-ordination and control of activities carried out by enterprise resources.* Without a description of such links it is not practical to (a) determine whether an enterprise is from a successful species or not or (b) determine how the behaviour of an enterprise and its component parts can be improved, whilst properly recognising its inherent capabilities, limitations and constraints;

(iii) as current trends, towards mass customisation, shorter product life times, globalised markets and rapid technological advance, induce faster rates of change a successful species of enterprise will need to learn from its experiences and adapt readily to new circumstances;

(iv) an enterprise will normally deal concurrently with the realisation and provision of a variety of products, services, contracts, etc. Each may relate to a different set of market and environmental conditions. Hence typically any given part of an enterprise may be changing at different rates and in different directions to other parts;

(v) contemporary enterprises are inherently resistant to certain types of change. Typical cycle times associated with significant organisational, social or technical system change will be of the order of many months to several years. Thus it follows that it is important to conceive and implement organisational and system structures which positively cater for and support change. However in any transition it is essential to recognise and handle issues associated with any legacy of outmoded human or technical system which needs to be retained.

The combination of modelling technology and infrastructure technology promises a way forward which ultimately may lead to the holistic conceptualisation and rapid incremental development of enterprises and their integrated manufacturing systems.

However, migration towards that situation will be constrained by the availability of suitable reference models and enterprise engineering methods which can promote and unify the deployment and development of modelling tools and infrastructural services. Also, notwithstanding such developments, it will be vital to seek ways of rapidly and effectively reusing or replacing legacy systems.

Bearing in mind the need and the current state of play in industry and academia this book by Valdew Singh provides a valuable text. It describes a pragmatic (but effective) way of combining the use of modelling tools and infrastructural services. As such it offers a reference model and method which can be followed by industrialist and advanced by academics. For industrialist the approach will facilitate the reuse and re-engineering of large, essentially monolithic, proprietary software components. What is more, it can lead to much improved alignment between systems operation and defined business goals. The approach was conceived by Valdew Singh as part of his PhD research but built on his previous extensive knowledge of solving industrial integration problems for Singaporean companies. Whereas for academics the approach embodies important concepts which can be extended and refined as new enabling technologies reach maturity. As such it can provide a valuable text for final year undergraduate and for postgraduate students.

Most importantly the reader should seek to understand the underlying concepts exemplified by Valdew Singh's approach to software interoperability. My own view is that they will equally apply post the year 2000, when business objects and software components libraries have become an industrial reality.


**Professor Richard Weston**
**Head of Manufacturing Systems Integration (MSI) Research Institute**
**Loughborough University, UK**
**January 1997**

# Preface

The introduction and adoption of computer-aided systems can indeed help rationalise and improve the productivity of manufacturing related activities. Such activities include product design, process planning and production management with CAD, CAPP and CAPM respectively. However, they tend to be domain-specific and would generally have been designed as stand-alone systems where there is a serious lack of consideration for integration requirements with other manufacturing activities outside the area of immediate concern. As a result, 'islands of computerisation' exist which exhibit deficiencies and constraints that inhibit or complicate subsequent interoperation among typical functional modules. As a result of these interoperability constraints, they typically yield sub-optimal benefits and do not promote synergy on an enterprise-wide basis.

Thus, Computer Integrated Manufacturing (CIM) with its underlying philosophy of traversing physical and functional boundaries to bridge the gaps among the islands of computerisation became the catch phrase of the '80s and the euphoria was immense. However, it is still a contentious issue as far as agreeing upon a universally accepted definition for CIM. Its scope varies in degree of importance and emphasis intrinsic to the perception held. Generally, it can be concerned with one or a combination of the following:

- computerisation of manufacturing related activities;
- systems integration to facilitate integrated information flow between upstream and downstream functional activities (i.e. engineering, production and management), and integrated material flow to support part manufacture;
- enhancement and optimisation of manufacturing processes.

Over the years there has been considerable discussion and debate about CIM. There are those who view it as a passing fad. There are others who see CIM as an evolutionary manufacturing philosophy of strategic importance which not only focuses on intra-organisation but also inter-organisation integration. The term 'Enterprise integration' is being adopted to reflect this wide scope of coverage. Arguably the use of the term CIM may be perceived to be rather outmoded, however, the author's choice to use the term has been a conscious decision taking into consideration its general acceptability and the understanding many have of it.

This book tries to explicate the issues concerning integrated manufacturing, with particular reference to application software interoperability. The enabling and emerging technologies and methodologies advocated are addressed. Prevailing gaps in technology and know-how are also discussed to have a better understanding of the shortcomings and limitations which can undermine the adoption and acceptance of integrated manufacturing and impose constraints when building such systems. A 'proof-of-concept' integrated manufacturing system is also discussed to show how such a system can be effectively realised.

**Valdew Singh**
**Singapore, 1997**
E-mail: valdew@pacific.net.sg

# Acknowledgements

The author wishes to take this opportunity to express his sincere thanks and heartfelt gratitude to:

- Professor Richard H. Weston from the Manufacturing Systems Integration Research Institute at Loughborough University of Technology (UK) and his team of researchers, particularly Paul Clements, Jack Gascoigne, Shaun Murgatroyd and Ian Coutts. This is in recognition of their invaluable contribution in terms of conceptual thinking and for some of the early foundation work on integrating infrastructure and application software interoperability which they had successfully established.

- Mr. Lin Cheng Ton, Principal and CEO of Nanyang Polytechnic, for his strong support and encouragement over the years;

- the team of research and development engineers from the CIM Centre, Nanyang Polytechnic, for their support, commitment and invaluable contribution of ideas towards the advancement of enabling and emerging CIM related technologies. Particular thanks are due to Lau Choon Mun and Zhang Binglu.

- all software and hardware manufacturers and vendors, particularly IBM, Hewlett-Packard, DEC, SUN Microsystems and ORACLE, for their invaluble technical contribution and support.

# Contents

*Preface*        ix

*Introduction*        1

**CHAPTER 1**        *Contemporary Solutions and Software Interoperability*        5

1.1 ISLANDS OF COMPUTERISATION        5
1.2 MANUFACTURING PARADIGM SHIFT        7
1.2.1 Integration and Software Interoperability        9
1.3 INTEGRATED MANUFACTURING SYSTEMS        10
1.4 SPECIFICATION FOR SOFTWARE INTEROPERABILITY        14

**CHAPTER 2**        *Current 'State-of-the-art' Scenario*        17

2.1 INTRODUCTION        17
2.2 INTERCONNECTION FACILITIES        18
2.2.1 'Pair-wise' Integration        18
2.2.2 Integrated Information Systems        20
2.2.3 Integrating Infrastructures        27
2.3 ENTERPRISE MODELLING        34
2.4 SYSTEM DESIGN AND DEVELOPMENT        36
2.4.1 Design and Modelling Methods to Support Life-Cycle Phases        37
2.4.2 Entry Point for System Life-Cycle Support        40
2.5 MEANS OF CONTROLLING SYSTEM BEHAVIOUR        41
2.5.1 Function and Information Entities Association        41
2.5.2 Model Enactment to Describe System Behaviour        44

**CHAPTER 3**        *Achieving Interoperability*        47

3.1 GENERAL CONSIDERATIONS        47
3.2 NUCLEUS FOR MANUFACTURING INFORMATION        48
3.2.1 Manufacturing Continuum Consideration        49
3.2.2 Manufacturing Methods and Information Requirements        53

3.3 INFORMATION MODELS   **53**
3.3.1 Specification   **54**
3.3.2 Characteristics of the Information Model   **62**
3.4 NEED FOR INTERCONNECTION AND INTEROPERATION   **64**
3.4.1 Requirements of Integrating Infrastructure   **66**
3.5 AN OVERVIEW OF THE METHODOLOGY DERIVED   **69**

**CHAPTER 4**      *Information Architecture*   *73*
4.1 MODELS FOR EXTENDED APPLICATION DOMAIN   **73**
4.2 CASE STUDY: APPLICATION OF REFERENCE MODELS   **73**
4.3 DESIGN CRITERIA FOR SYSTEM-WIDE DATA REPOSITORY   **78**
4.3.1 A Logical Database Model of the Data Repository   **81**
4.3.2 Future of Relational Database Management Systems   **84**
4.3.3 Database Access Approach   **85**
4.3.4 Database 'Driver' Requirement for IIS   **87**
4.4 SUMMARY   **89**

**CHAPTER 5**      *Integrating Infrastructure to Underpin Interoperation*   *91*
5.1 FUNCTIONAL INTERACTION REQUIREMENTS   **91**
5.2 CONTEMPORARY SOLUTION: FUNCTIONAL INTERACTION   **93**
5.2.1 Overview of Systems Integration Manager   **94**
5.2.2 Focus and Limitations   **96**
5.3 FUNCTIONAL INTERACTION MANAGEMENT SERVICES   **98**
5.3.1 Distributed Functional Interaction Management   **105**
5.4 REQUIREMENT FOR FRONT-END USER INTERFACE   **106**
5.5 SUMMARY   **108**

**CHAPTER 6**      *System Life-cycle Support*   *111*
6.1 REQUIREMENT FOR INTEGRATED LIFE -CYCLE SUPPORT   **111**
6.2 CASE STUDY: REALISING THE INTEGRATIVE APPROACH   **112**
6.2.1 Information Model Enactment   **115**
6.2.2 Functional Activity-based Modelling   **118**
6.2.3 System Behaviour Enactment   **119**
6.3 SUMMARY   **125**

**CHAPTER 7**   *Use and Appraisal of the Methodology Derived*   *127*

7.1 INTRODUCTION   127
7.2 PROOF-OF-CONCEPT IMPLEMENTATION   129
7.2.1 Implementation Steps   132
7.2.2 Analysis and Discussion   142

**CHAPTER 8**   *Conclusion*   *145*

8.1 GAINING A BROAD PERSPECTIVE   145

*Appendix I*        151

*Appendix II*       161

*Appendix III*      165

*Appendix IV*       169

*Glossary*          175

*References*        181

*Index*             197

# Introduction

Admittedly, the market for books on CIM might be well supplied, but the approach taken in this text is surely a balanced one where the enabling and emerging technologies are discussed in view of the industry's immediate concerns and pragmatic requirements. This book is written with a broad spectrum of people in mind, particularly the systems integrator, industrial end users and academics. Here the underlying concepts, methodologies, software toolset, and enabling mechanisms advanced by the author to realise integrated systems are covered.

Such issues concerned with facilitating software interoperability and overcoming certain limitations and restrictions found when achieving applications integration in contemporary forms of functional modules have been considered in Chapters 1 to 5. In Chapter 6 the notion of an integrative 'model driven' methodology is introduced which has been derived to formally structure and support change management for system life-cycle support. Here the concept of 'enacting' function and information models is examined to provide consistent interface and support over the system life-cycle phases. Chapter 7 focuses on the implementation of a proof-of-concept system to illustrate the application of the methodology adopted and developed by the author to enable interoperation of functional components.

*Chapter 1:* The need for software interoperability and the issues which affect integration between software applications are discussed. Some of the major considerations are examined to develop pragmatic and viable interoperable systems. Also outlined are the nature and scope of software interoperability, and a requirement specification to enable interoperation among functional components.

*Chapter 2:* Inherent constraints and limitations which either inhibit or complicate interoperation among contemporary forms of computer-aided solutions are identified and examined. A review of contemporary approaches and methodologies available to tackle particular aspects of the interoperability problem (i.e. interconnection, system design and development, reference models and system behavioural issues) is conducted. Based on the review, some of the outstanding issues and 'gaps' in knowledge are highlighted.

***Chapter 3:*** The need for a reference information model, which can be widely applicable across the manufacturing continuum to promote interoperability among functional activities, is highlighted. The components and functionality of an integrating infrastructure (IIS) to enable interconnection among functional components are described. Also provided is an overview of the methodology advocated by the author to enable software interoperability.

***Chapter 4:*** An information architecture which can support software interoperability is discussed. It includes the specification of reference models which describe information entities of common interest shared among production planning, product design, process planning, finite capacity scheduling and cell control systems. Included is a case study carried out in collaboration with the University of Bradford Management Centre, UK, to demonstrate the application of the generic reference models identified and described. The design criteria for the system data repository are specified. The mechanisms (such as the database 'driver') to enable open information access in a consistent and reliable manner from the system data repository are also described.

***Chapter 5:*** The emphasis in this chapter is on system behavioural issues. The methodology adopted, mechanisms developed and system management tools required to formally structure and facilitate functional interaction (among distributed interoperating functional components during run-time) are discussed. The need for functional interaction management services, which has been conceived as part of the overall methodology to enable effective functional interaction, is also explained. Its functional capabilities and level of effectiveness to facilitate interaction is compared with a contemporary commercial software application, namely Systems Integration Manager, which is designed to tie together a disparate set of software applications into a coherent system.

***Chapter 6:*** An integrative 'model driven' methodology advocated by the author to provide life-cycle support of integrated manufacturing systems is highlighted. The software toolset developed which is coupled closely with $IDEF_0$, $IDEF_{1X}$ and EXPRESS based modelling tools, to formalise and support implementation, run-time and change processes is described. The exploitation and enactment of activity-based functional and information models by the software toolset to structure downstream life-cycle processes is also covered.

***Chapter 7:*** In this chapter a proof of concept implementation study has been carried out in order to illustrate the application and ascertain the level of effectiveness of the methodology described, which seeks to enable interoperability among functional components.

***Chapter 8:*** A summary of the methodology described is provided. Future areas of development expected to have an impact on enhancing the software interoperability methodology described in the book are also highlighted.

# 1 Contemporary Solutions and Software Interoperability

*He who does not think far ahead is certain to meet troubles close at hand.*
Confucius, Born 551 B.C.

## 1.1 ISLANDS OF COMPUTERISATION

In recent years the proliferation of information technology and the increased affordability of computer hardware have contributed to the increasing use of computers in manufacturing. However, islands of computerisation are generated. This phenomenon can in part be attributed to the dominant influence of Taylorism, which places emphasis on specialisation and distinct division of responsibility. Traditionally, this has led to a compartmentalisation of business, engineering and production activities, all of which are required to support the product life-cycle. Consequently, there has been a proliferation of computer-aided systems where each has been designed with Tayloristic principles in mind to address some focused aspect of the

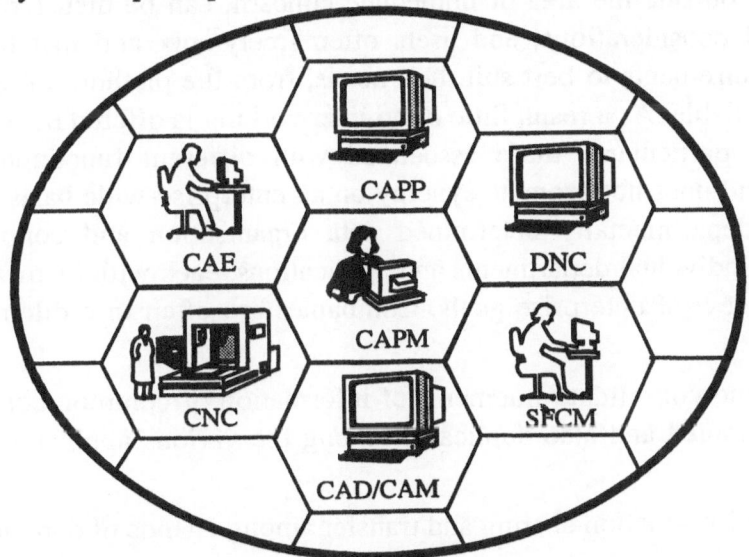

**Figure 1-1** Insularity of contemporary solutions
*Specialised domain-specific applications which exhibit functionality for the rationalisation, improvement, control and execution of activities in support of part manufacture.*

manufacturing domain in order to facilitate product introduction, planning processes and enable the control of manufacturing activities and processes on the shop floor. They are implemented in a variety of forms and serve different purposes, many of which are typically classified under the general headings of Production Planning and Control (PPC), Computer-aided Design/Manufacture (CAD/CAM), Computer-aided Engineering (CAE), Computer-aided Process Planning (CAPP), Computer-aided Quality Control (CAQC), Distributed Numerical Control (DNC), Shop Floor Control and Monitoring (SFCM), etc. As illustrated in Figure 1-1, normally, they are implemented in the form of stand-alone software packages and systems, each focused on enabling localised efficiency and productivity improvements with respect to different aspects of a manufacturing enterprise. The adoption of these computer-aided solutions would, to a certain extent, rationalise and streamline the functional activities concerned.

However, computer-aided solutions of this type are conceived, implemented and supplied from various sources, leading to significant heterogeneity in terms of the computer hardware, systems software, systems management and communications systems they employ. Very often this is accentuated by the fact that they are conceived and implemented in the absence of an overall company-wide strategic plan of how they should be implemented and how the integration requirements of other software applications, outside the area of immediate concern, can be met. Driven by myopic departmental considerations, end users often freely mix and match hardware and software requirements to best suit their needs, from the plethora of systems that are currently available. As a result little or no interworking is offered between constituent applications, particularly those associated with different functional areas of the enterprise, and does not promote synergy on an enterprise-wide basis. This results in specialised departmentally determined data organisation and contributes to data hoarding by individual departments and applications, each with its own restricted and incomplete view of enterprise goals. Companies are often in a dilemma when they need to:

- accrue and consolidate fragments of information of common concern which are distributed and also duplicated among the various application software; and

- facilitate information sharing and transfer among islands of computerisation.

This can be attributed mainly to the insular nature of contemporary computer-aided solutions where their interconnections (to facilitate transfer and sharing of information) are severely constrained by their proprietary nature and the level of heterogeneity exhibited by them. Hence there is much reliance on informal proprietary links to (a) convey messages and information, and (b) co-ordinate transactions between specialised departments and applications. Consequently, significant delays and errors in transactions normally occur, thereby promoting opportunities for misunderstandings and conflicts. This undoubtedly hampers the productivity and operational efficiency of the enterprise concerned.

## 1.2 MANUFACTURING PARADIGM SHIFT

The demands imposed by today's highly competitive and rapidly changing customer-oriented market have resulted in a paradigm shift for manufacturing suggested by the emergence of the extended and integrated enterprise. No longer can manufacturing enterprises afford to perform their various domain-specific functional activities in complete isolation and in a disjointed manner; particularly in this era of time-based competition where the drive is towards being responsive to customers' demands and changes in other market forces so as to exploit market potential and capitalise on existing business opportunities at an early stage.

Thus, there is increased demand for a degree of interworking among functional modules to enable the manufacturing enterprise to function as a synergistic whole — that the normally separate parts of an enterprise could share common resources and goals. This would help (a) improve responsiveness, (b) avoid errors and delays in and among transactions, and (c) alleviate potential conflicts and contentious situations by providing:

- accessible, coherent and up-to-date view of information of common concern to support decision-making in a more timely and accurate manner; and

- a global perspective of the work domain, so as to have a better appreciation and understanding of its impact on and close association with other activities in the enterprise.

It would no longer be necessary to expect a single software application to fulfil its purpose without support or reference to data and events which are handled by other closely related application systems. This requires application software interoperability

**Figure 1-2** Interoperation among domain-specific functional modules

**Figure 1-3** Integration levels in the manufacturing enterprise

where the various islands of computerisation are linked to facilitate interaction and sharing of commonly used engineering, manufacturing and management information (see Figure 1-2). Drucker (1991), in his assessment of the factory of the future, shares this view where he emphasised that:

*The factory of the future will be an information network. Sectors and departments will have to think through what information they owe to whom and what information they need from whom. A good deal of this information will flow sideways and across departmental lines, not upstairs as with traditional plant ...*

## 1.2.1 Integration and Software Interoperability

The process of systems integration is wide ranging and holistic in nature. As illustrated in Figure 1-3, the AMICE CIM-OSA consortium classified integration in the manufacturing enterprise within the following three levels (ESPRIT 1989, CIM-OSA 1989):

- **Business Integration** At this level, enterprise goals and strategic business issues are considered.

- **Application Integration** Integration at this level is defined as concerning interoperation between applications to facilitate data sharing and information exchange.

- **Physical Integration** Physical integration is mainly concerned with data and inter-process communication issues. It expects this level of integration to be provided by current information technology concepts and standards.

Within the context of integrated manufacturing, the term 'software interoperability' has been used to imply the ability of separate software applications to functionally interact to meet collective goals. Bearing in mind the classifications, software interoperability is aimed at the physical and application integration levels of enterprise-wide integration.

Application software interoperability is widely conceived as requiring data integration as well as functional integration with the resultant effect of linking and synchronising the behaviour of processes in different subsystems of an enterprise. The underlying interaction processes will involve an exchange of messages and the sharing of information of common interest among a group of software applications so that the applications behave (both individually and collectively) in an effective manner whilst realising system-wide goals.

When dealing with the heterogeneous and diverse environment, standardisation is essential to help improve compatibility, interoperability, interconnectivity and longevity regarding those discrete software and hardware entities incorporated. Over the past decade significant progress has been made towards improving the 'hardware portability' of manufacturing related application software building blocks where software vendors have sought to adopt the use of de jure and de facto computer networks, operating systems, database, fourth generation languages and graphical user interface standards (see Figure 1-4). This trend towards hardware portability has enabled the functionality implemented by any particular piece of application software to be separated from (a) the computer hardware on which it is run, and (b) the data on which it operates. This alleviates certain problems associated with installing, using and changing individual application software building blocks.

However, in seeking significant enhancement at the level of interoperability achieved between chosen software building blocks (where information exchange is a key requirement) common function and information models, which encapsulate key general attributions of various forms of manufacturing application software, are also required. Currently, there is an absence of such models.

## 1.3 INTEGRATED MANUFACTURING SYSTEMS

Although many are generally aware and acceptable to the notion of integrated manufacturing there is still a lack of proper understanding of its broader implications. It is commonly construed as merely about physically interlinking a set of computer hardware and application software to support part manufacture. This represents only one of its primary concerns. Integrated manufacturing should be perceived as a manufacturing philosophy geared towards:

- promoting inter- and intra-organisation integration by facilitating close co-ordination of associated functional activities to enhance the productivity and efficiency of the manufacturing enterprise; and

- consolidating and sharing of information of common interest so as to maximise its value to support decision-making and part manufacture.

| Requirements | Standards | | Functions |
|---|---|---|---|
| | De jure | De facto | |
| Windowing | X-Window System OSF/Motif | | Consistent User Interface |
| Graphics | GKS-3D PHIGS IGES | | |
| Operating System Interface | (IEEE POSIX 1.003.1, Emerging 1.003.n) CORBA | | Access to System Services |
| Database Definition/Access | SQL, Emerging SQL Access and Remote Data Access (RDA), Integrated Database Application Programming (IDAP) | | Information and Resource Sharing |
| Enterprise Repository | Emerging IRDS, ATIS | | |
| File Sharing | Network File System (NFS) Emerging OSF DCE AFS | TCP/IP | |
| Mail | X.400 | | Enterprise Communications |
| EDI | ANSI X.12, EDIFACT | EDIF | |
| RPC | Emerging OSF DCE | | |
| Plant Floor Communications | MMS | RS-232 | |

**Figure 1-4** The standards continuum and a categorisation of their purpose

## (I) Inter- and Intra-organisation Integration

Integrated manufacturing permeates across every facet of the enterprise to realise synergy. Conventional physical and functional boundaries among domain-specific functions, which span engineering, production and management activities and systems, need to be traversed to enable close communication and the coherent and integrated flow of information of common interest within the organisation and externally, such as with vendors and suppliers who may be part of the supply and distribution chain. The aim is to bring together the various sub-entities of the manufacturing enterprise to enable it to operate as a whole and to help facilitate co-operative and 'more-informed' decision-making to fulfil its required business needs.

## (II) Maximising the Value of Data

This is a key strategic challenge as information is identified now to be the most important asset in any organisation. In integrated manufacturing there is a need to facilitate distributed information access, manipulation and management. Very often information of common interest, as generated by the various upstream and downstream activities, needs to be extracted and consolidated so that it can be easily shared and transparently accessed. There is also increased realisation that the best way to reference shared information is to consolidate it in a single, logical store or more commonly referred to as a data warehouse. Boundaries between domain-specific functional activities can to some extent be overcome.

The choice and type of information, however, would vary depending on the functional modules that need to be interconnected to facilitate interoperation. For example, production planning and control and the following three areas are closely associated. They share common information entities but are viewed from different perspectives:

- **Process Planning** This serves as a technological bridge between engineering and manufacturing and provides a blueprint for part manufacture. It defines the sequence of operations required to manufacture a product and selects the manufacturing resources to carry out material processing. It also incorporates data concerning manufacturing parameters. According to a report by Halevi and Weil (1992) process planning systems should be able to communicate with other company functions, especially production planning and product design, in order to achieve significant benefits in co-ordinating manufacturing activities.

- **Product Design** When designing a product for manufacture or carrying out design modifications input on availability of manufacturing resources (including machines, toolings, fixtures) and capability are required. Such information can be gathered from production planning and control and process planning. Availability of this information to support decision-making would help reduce design-to-manufacture lead time and avoid unrealistic product specifications which might cause serious problems during actual part manufacture.

- **Shop Floor Control and Monitoring** Shop and cell control systems have responsibilities for a segment of the shop floor and are required to despatch planned orders (which are generated from production planning systems), co-ordinate, control and monitor the operation of the components of a shop or cell. They will also be responsible for shop floor data acquisition, thereby enabling production status feedback to the production planning and control system.

End users today are becoming more discerning when choosing solutions for implementation as they view long-term benefits important in terms of safeguarding their investments. As a result there is increasing demand for technology-independent solutions; ease of development and management of change are essential to support the system through its useful life-cycle. And integrated manufacturing systems are no exception to the rule. In order for integrated manufacturing systems to gain wider acceptance the following need to be addressed:

- **Coping with Legacy Systems** This refers to a previously installed base of systems, components and software. Integrating legacy elements always pose a difficult and costly problem because the elements do not normally conform to the methods and standards which will be adopted in current generation solutions. They are in danger of becoming technically obsolete due to rapid changes in technology. When obsolescence occurs, one of the following actions will normally be required:

  (a) modification or enhancement in functionality so as to upgrade and make functionally effective again;

  (b) retrieval of existing information which can be a vital resource in support of the operation of the manufacturing enterprise; and

  (c) discarding entirely and replacing with a viable alternative when there is neither any chance nor need for (a) and (b).

With a legacy software, the possibility of performing (a) may be remote. Often this is due to a lack of proprietary knowledge, support and expertise to carry out necessary changes (such as amendments to the required source programs), thereby making the task arduous indeed. Furthermore, it is uncommon for software manufacturers to reveal and release information pertaining to the source programs of their developed applications. Understandably, this is to protect their vested commercial interest and rights to intellectual property embodied in their products.

Hence in seeking to facilitate integration, generally speaking (b) seems to be a more workable and pragmatic means of dealing with legacy software than (a). However, in order to achieve an acceptable level of interoperability via information sharing, often the following prerequisites are essential:

- the information source (which forms part of the legacy element) can be independently accessed; and

- the information architecture and schema used by the legacy component are clearly understood in terms of their structure and composition.

Thus as an initial step, there is an important need to define the nature and form of a suitable information model which can be used to help facilitate data exchange among functional modules. Such a model will need to describe information of common interest to the typical functional modules.

- **Change Management** Contemporary computer-aided systems are normally designed and implemented to address a set of problems prevalent in current company situations. Once installed, inevitably any solution is at risk of becoming obsolete if it is incapable of coping with situations other than those for which it has been designed. In fact system extensions are the most commonplace of software activities and the most expensive driver of software costs (Horowitz 1993). Thus, a more formal and structured methodology for system design, implementation and maintenance is required to provide 'open' access and ease of management change.This would effectively offer a migration path to reflect changing business needs.

# 1.4 SPECIFICATION FOR SOFTWARE INTEROPERABILITY

It can be stated that ideally application software interoperability should imply an uninhibited functional interaction and intercommunication amongst functional components through a free exchange of shared information. This also implies a need to standardise the interfaces between the software components, especially concerning the control of information exchange among the components. Therefore, it is clear that the concept of software interoperability extends beyond that of software portability, i.e. interoperation is required over different hardware platforms, operating systems and information access and storage systems.

To summarise, the following requirements need to be satisfied to enable software interoperability in an effective manner to overcome (a) limitations inherent in contemporary application software and solutions, and (b) associated and inherited difficulties and problems involved in achieving their interoperation:

**Information Sharing Requirements**
- Information models which describe information of common concern to various interoperable functional modules thus effectively serving as a precursor to their interworking.

- An information architecture which establishes structure and uniformity whilst enabling information access and sharing among the functional modules.

**Interconnection Facilities**
- An integrating infrastructure which simplifies and structures interconnection by (a) separating integration and application issues, (b) providing inter-process communication services, and (c) mapping of distributed processes (embodied in the functional modules) on the physical resources contained within a target manufacturing system.

**Functional Interaction**
- Capability for controlling and co-ordinating the sequence of (run-time) activities in a data- and event-driven manner, based on their dependencies, information needs and availability.

**System Design and Development Capability**
- Provision for a more formal and structured approach to:
    - engineering integrated manufacturing solutions; and
    - supporting them over their useful life.

This is to facilitate ease of development and change management in order to be adaptable and responsive.

# 2 Current 'State-of-the-art' Scenario

## 2.1 INTRODUCTION

Very often when the process of systems integration is carried out in the absence of an overall strategy and clarity in the intrinsic requirements to satisfy end user and business needs it becomes a thing of chance and adventure rather than a regularly understood business. This would give rise to inherent constraints and limitations which contribute the following:

- monolithic, vertically integrated applications
- closed proprietary systems
- no discernible software architecture
- expensive maintenance and evolution.

This can be generally attributed to the lack of architectural focus and informal *ad hoc* programming practice in developing bespoke and proprietary solutions which inhibit system extension. In fact system extensions are the most commonplace of software activities and the most expensive driver of software costs.

There are various methods used which attempt to resolve and overcome (a) certain limitations inherent in contemporary solutions, and (b) difficulties and problems associated with achieving a degree of software interoperability. In this chapter, the nature and status of these methods are discussed and the chapter is structured with reference to certain aspects of the requirements specification identified in Section 1.4, namely with respect to commonly used ways of providing:

- interconnection facilities
- information reference models
- system design and development capabilities
- means of controlling system behaviour.

## 2.2 INTERCONNECTION FACILITIES

Interconnection can be viewed as establishing electronic data interchange between the various islands of computerisation. Here components of the functional modules are interconnected to provide a low level data inter-communication and information transfer facility. Three broad classes of approach can be identified which will herein be referred to as:

- 'pair-wise' integration
- integrated information systems
- integrating infrastructure.

### 2.2.1 'Pair-wise' Integration

The 'pair-wise' integration approach is frequently adopted to interconnect the functional modules, thereby enabling information transfer between software applications (or components) of such systems. This approach can be characterised as follows:

- **Requires the Development of Bespoke Interfaces** (see Figure 2-1) These interfaces will normally need to be custom designed (each at a relatively high cost) to realise a level of integration between interoperating pairs of application software. The complexity of such systems will grow substantially (theoretically in a square law fashion) as the number of interconnected applications grows. This implies that for a set of $n$ different systems which need to be linked, $n(n-1)$ different interfaces may need to be developed; what is worse is that potentially $2(n-1)$ interfaces need to be adapted whenever a single application system is changed.

Typically this approach results in the incorporation of knowledge (concerning the need to interoperate) into individual application software and its associated 'drivers'. This will include knowledge of other application software, data sources, data access mechanisms, communication protocols, communication channels, data formats and data structures. This results in inflexible, application-specific and rigid solutions which can be classified as 'hard' integration. Such solutions will not be easily supported (in terms of available technical expertise) and the understandable reluctance of vendors to release detailed product specifications (as this may embody knowledge which provides them with a competitive advantage) will often lead to

**Figure 2-1** Interconnecting software applications through 'pair-wise' integration

**Figure 2-2** 'Pair-wise' integration via import/export filters

sub-optimal interworking between components. Furthermore, the cost, in terms of resources and time, of subsequent modification may be so great as to render the solution obsolete as soon as requirements change significantly.

- **Utilisation of Import and Export Filters** For some software packages, export filters are provided as a built-in utility to enable the user to have independent access to its proprietary data. The filters assume responsibility for pre- and post-processing of data which is specifically selected by the software manufacturer to be made available to the user. Here some restrictions are necessary in order to maintain data security and integrity through controlled access. The data will be automatically converted, via the filters (see Figure 2-2), to conform with a required database or file format supported by the software package. Generally the file formats adopted here will either be:

  - a compatible format to enable direct data transfer between software packages; or

  - a neutral format, such as in the form of a flat file, to allow intermediate data transfer (in the absence of any compatible database or file format) between software packages. In this case, further overhead processing will be required to retrieve, manipulate and store the relevant data.

  Data conversion and data transfer are normally performed in a batch mode processing.

  The main limitations of using filters are (a) that they provide only restricted access to selected data, and (b) the details of data format and structure can be lost in the transfer process.

With such contemporary 'pair-wise' methods of interconnection, each software application manages its own data and this can result in significant access times and data transfer times as a result of the inherent mechanisms used to retrieve and make data available to other applications. As a result of such delays there is no guarantee that data will be sufficiently up-to-date to support other dependent applications.

## 2.2.2 Integrated Information Systems

There is a growing emphasis on the development of integrated information systems based on a data integration approach. This approach can help prevent cumbersome information transfer (as illustrated in Figure 2-1) and can also reduce delays in information transfer times. The underlying principle of integrated information systems is that they consolidate information into a central data repository (or common pool) by seeking to closely map common data to be exchanged and shared among applications into that pool. Generally speaking, each application is required to support a capability to export and import schema to this data repository for which a global schema of common data models is defined (refer to Figure 2-3). Potentially, this approach should enable information of common concern (which typically will accrue at one stage in the production chain) to be included in the data repository, thereby making it accessible to application software used at other stages in the chain.

**Figure 2-3** Interconnecting software applications through integrated database

This approach has been realised industrially (at least to some extent) in recent years and has led to a degree of rationalisation within divisions of some enterprises. However, as described below major practical problems remain with regard to the development, enhancement and maintenance of integrated manufacturing systems which impede the widespread adoption of integrated information systems and may generally restrict interoperation among functions modules. Many of these problems arise from:

- **The tight coupling that normally exists between functions and their associated information.** This (a) makes information access difficult or even impossible, and (b) leads to unintended propagation of the effect of changes made, i.e. change to individual applications (comprising software processes and their associated systems) can have significant effect on the operation of other applications.

- **A lack of adherence to standard architectural models of functionality and information.** Rather, contemporary software applications are designed using proprietary models of function and information which are determined by the manufacturer. The main disadvantage is that information of common concern to software applications is often duplicated, translated, and re-interpreted by different software applications. This gives rise to problems of data integrity and consistency as well as significant database management problems. Furthermore, semantic integrity has to be ensured and maintained between valid combinations of data items fragmented across various databases.

- **Heterogeneity in database systems, particularly with regards to their logical data model.** As elaborated, the following three logical data models most commonly supported by database management systems are:

  (a) Hierarchical model
      Data is represented in a hierarchical or tree structure. Tree structures provide a natural way of modelling truly hierarchical real world relationships where one-to-many segment types can be defined to represent successive levels in a tree structure to relate entities to one another. However, in many situations relationships do not naturally fit into this model. For instance it is not easy to directly represent relationships between segment types at the same level in the hierarchy, nor is it possible without introducing data duplication to represent many-to-many relationships between entities.

  (b) Network model
      In the network model data is represented in a network (or plex) structure

where any node can be connected to any other node represented in the structure. Network structures offer a greater scope to represent data relationships than hierarchical structures, albeit at the expense of simplicity (at least with respect to physical storage structure). The need to transform many-to-many relationships by the construction of a network model does mean than more or less irreversible decisions have to be made about the nature of the relationships between entities when the data model is designed. It should be noted that the network model, whilst permitting a representation of many-to-many relationships without introducing duplication of the duplicating record occurrences, does make retrieval of data a laborious process.

(c) Relational model

In relational model entities, relationships and attributes are represented in the form of two-dimensional tables known as relations. Records are assimilated to the rows of the table and each set of attributes forms a column. In a relational database entities are stored totally independently. Logical associations among the stored data are exploited through relational operations, such as select, project and join which can be used to create new tables. The application of any (relational) operation produces an object which is itself a relation (which can be stored as a new table in the database). Thus any number of operators and relations can be combined in a 'relational expression' and used to answer almost any query. The use of the relational model rather than hierarchical or network models is seen to demand less compromise in transforming the real-world model of the conceptual data model, although the processing overhead it requires is still often a serious deterrent to its use for many applications.

Thus there are major difficulties involved in attempting to interconnect these heterogeneous database systems. This is primarily due to non-uniformity in heterogeneous database management systems and physical storage structures where there are serious concurrency problems related to transactions and controlled data access. In addition, the following must also be reconciled:

• differences in database schema; and
• semantic differences among data items.

However, with the advent of Distributed Computing Environment (DCE) technology it is possible for the functionality of applications to be divided and made readily accessible to specific users based on the client/server model so as to facilitate on-line transaction processing. Adopting DCE technology (as elaborated next) for developing client/server applications would provide the following advantages but it does not cater for integration and interoperation with other functional modules:

- portability and low level interoperability over a range of computers and networks; and

- ability to share data and services efficiently and securely regardless of the number of computers used or where they are located.

## DCE Technology

Traditionally, the client/server model consists of a user interface (client) and a database (server) which is known as the two-tier client/server architecture. The clients represent the user and the servers provide the services requested by users. In addition to the user interface and database, applications representing and supporting specific functional activities have to be developed within the chosen architecture to filter, manipulate, format and present the data generated. These 'business applications' become part of either the client (Figure 2-4) or the server (Figure 2-5).

However, in a dynamic and complex manufacturing environment where there is a constant need to evolve and adapt to changes in business needs to keep pace, it is quickly realised that the two-tier client/server architecture can no longer satisfy changing requirements from the perspective of cost, maintainability or performance. Subsequently, to allow for greater flexibility, we now see a new approach being adopted which separates the business applications into their own architectural tier. The resultant three-tier client/server architecture has, therefore, the traditional user interface clients, data servers and an independent functional layer of business applications (Figure 2-6). The business applications can be easily linked to the clients and server through the incorporation of standard Application Programming Interfaces (APIs) and network protocols made available.

The Remote Procedure Call (RPC) facility is the basis for all DCE client/server communications and therefore is fundamental to the distribution of services in DCE applications. The RPC mechanism enables a procedure invoked by one process (the client) to be executed, possibly on a remote host, by another process (the server). The client and server hosts do not need to have the same operating system or the same hardware architecture. DCE RPC conforms to a set of specifications collectively

Source: The Standish Group International, *'Open OLTP Report'*, 1993

**Figure 2-4** Remote Data Access (RDA) method

*The RDA method maintains one central repository for data and attaches business applications to the user interface. This is relatively easy to develop and deploy, offering a wide variety of off-the-shelf development tools. However, in a dynamic manufacturing environment, these systems are difficult to maintain since a copy of each business application exists for each client. Because they are constantly retrieving data from the remote data repository, applications operating in this architecture tend to be network intensive.*

**Figure 2-5** Data Base Server (DBS) method

*The DBS method isolates the user interface, allowing for lower cost and greater flexibility in choosing user interface hardware resources. Data integrity improves by sharing application services. Consolidation of application services would seem to simply system maintenance, but in reality, a combination database and application tends to become monolithic over time, and as a result, grows more difficult to maintain. This monolithic application also experiences performance bottlenecks in trying to keep up with requests from an ever-increasing number of clients.*

**Figure 2-6** Three-tier client/server architecture

**Figure 2-7** Interconnection via an integrating infrastructure

known as the Network Computing Architecture (NCA). The NCA specifications define the protocols that govern the interaction of clients and servers, the packet format in which RPC data transmitted over the network, and the Interface Description Language (IDL) that is used to specify RPC interfaces.

The Open Software Foundation (OSF) is actively working towards fulfilling the need for a standardised approach to creating and executing secure client/server applications in complex, highly networked environments.

## 2.2.3 Integrating Infrastructures

Integrating infrastructures (IIS) are systems integration enabling tools that allow applications to 'functionally interact'. Their main purpose is to structure, service and, where possible, simplify interconnection between the component elements of software systems. Increasingly there are a number of such systems integration tools appearing on the market, namely PlantWorks/DAE (Distributed Automation Edition), WorkStream/BaseStar and Industrial Precision Tools (IPTs) from IBM, Digital and Hewlett Packard respectively, which serve as the backbone for enterprise-wide integration.

As illustrated in Figure 2-7, an IIS can be charged with resolving differences in a physical system relating to heterogeneity, distribution and data fragmentation. It can assume responsibility for maintaining a knowledge of integration details (such as the networks used, the hardware and operating systems that software components are run on, the location of an information fragment, etc.) so that software components themselves need only have knowledge of how to use the IIS (i.e. not of each other).

An IIS is usually supported by software tools to help alleviate the complexities inherent in most systems integration projects. A range of development tools, better known as middleware typically in the form of APIs, for programmers and third party developers can be offered and may provide:

- consistent user and device interfaces to allow interaction over the IIS; and
- structured access to common integration services for (a) inter-process communication, and (b) information sharing and management via a data repository.

**Figure 2-8** Middleware functionality: The API perspective
*The API perspective portrays each major interaction paradigm as a separate technology with little relation to the others. the taller the bar, the higher is the level of the API. All middleware presents higher level APIs, some only to handle differences between network transport APIs, but most provide services beyond that. The gap between a middleware bar and the applications on top signifies the relative work required for the application to interact with the middleware.*

The following categories of middleware services are available (see Figures 2-8 and 2-9) depending upon the nature of the requirement:

• **Message-oriented Middleware (MOM)** MOM systems offer a very basic set of commands — often as few as SEND and RECEIVE. It lets programs send data to programs in real time (or slower). MOM is as general purpose as it can get for the programmer. Application developers create application-specific functions or routines built on these basic messaging functions. As far as transports go, the interface is even simpler than programming to a particular network transport API, like TCP/IP sockets. Many MOM products exist, including Covia Communications Integrator (CI), PeerLogic Pipes, Momentum XIPC/Message Express, IBM MQSeries, Digital DECmessageQ and NetWeave. Relational Database Management System (RDBMS) vendors like ORACLE and Sybase are also getting into the fray as well with products like ORACLE Mobile Agents and EMS.

**Figure 2-9** Middleware API architectures

- **Remote Procedure Call** RPC is function-oriented and is used in a client/server computing environment. Developers define their own application-specific functions using an Interface Description Language (IDL), and then compile that function into the client and server stub code that actually does the networking. The application just makes normal function calls. Developers essentially create their own APIs. Many RPC-based products only generate function stubs for 3GL languages like C, but some support 4GL products like PowerBuilder. When distributing existing host-based multiuser applications, the RPC approach is very intuitive: each existing function can be split across the network as needed — just recompile to distribute a given function. RPC-based products are fairly automatic compared with message-oriented solutions. RPC-based products come from many vendors, which includes DCE standard vendors such as IBM, DEC, SUN, Hewlett-Packard and many others.

- **Data Access** Data access products offer data-oriented APIs that reveal data in tables. Using a standard API, which is Open Database Connectivity[1] (ODBC) compliant, applications get at remote data using Structured Query Language (SQL). However, most RDBMS vendors have proprietary APIs in addition to ODBC, since those let them directly expose unique RDBMS functionality. If the application only needs to have its data distributed to, and shared from, database servers, data access works well. Many tools already support this API, and since no application-specific programming is required on the back end, it empowers *ad hoc* application development. Some of the data access products available commercially include SQL*Net, Open Client/Open Server and DB-Library from ORACLE, Sybase and Microsoft respectively.

- **Distributed Transaction Processing (DTP)** DTP monitor products offer a middleware environment oriented toward handling transactions over a network. Transaction monitors add BEGIN and END TRANSACTION semantics to the generic SEND and RECEIVE commands. By using DTP services, the application doesn't have to include logic to assure transaction integrity. DTP products are often built on top of message-oriented or RPC-based technology but add significant control and management functionality. Most DTP vendors are working to implement new X/Open standard APIs, but today most have proprietary APIs. DTP products interact with database resources via the X/A standard interfaces supported by most RDBMS vendors. Products include Novell Tuxedo, Transarc Encina, AT&T GIS Top End and IBM CICS.

- **Object Request Broker (ORB)** The adoption of ORB would require the fragmentation of software applications into appropriate atomic level of base objects which would encapsulate specific functions. The objects communicate via messages and can be created using an object-oriented software, for example C++, and they can be directly incorporated to the ORB middleware. An ORB allows the developer of an application to generate all the objects necessary and to provide the ability to link independent applications, exchange data between them and to enable them to invoke functions on each other. This is similar to the RPC-based approach, except that it is object-oriented rather that function-oriented in nature. The principal benefits include sensible use of resources, faster development, high degree of fault tolerance, and improved flexibility and reusability of software components on a

---

1. **ODBC** allows developers to design applications based on a common API rather than a specific DBMS, thus creating applications that are easily portable from one DBMS to another.

comprehensive scale.

If the application developer does not use an ORB, the programming job will be complicated by having to create the links between objects and the functions or other APIs offered by other middleware solutions. The ORB world has many standards: the Object Management Group's Common Object Request Broker's Architecture (CORBA[2]), along with de facto standards like Microsoft's Object Linking and Embedding (OLE) technology and OpenDOC architecture.

CORBA supports distributed heterogeneous computing but much of it currently defines developer APIs, although later versions are expected to define services (via the APIs as well). OLE technology is, however, aimed at the top end of the application interaction hierarchy where OLE-enabled applications are allowed to interwork directly in the manner of objects through automatic initiation and passing of messages. OpenDOC is a vendor-neutral alternative to Microsoft's OLE technology. It aims to see in an era of component software where users create their own bespoke applications by interconnecting mini applications (or components) dedicated to specific functions. The OpenDOC compound document architecture is from CI Labs[3] and it uses System Object Model (SOM) which is IBM's implementation of the CORBA specification. Applications developed based on OpenDOC would work seamlessly together to create the appearance of a single application. Thus software components which are OpenDOC-compliant will be able to interoperate across different platforms, and, for added good measure, with OLE- and CORBA-compliant objects.

ORBs are commercially available from a number of system vendors (such as IBM SOM/DSOM, HP, Digital, NEC, AT&T GIS) and many third parties, including Iona (Orbix), Expersoft (PowerBroker), PostModern (ORBeline) and NeXT.

Increasingly, the trend is towards offering multiple middleware products layered together as a suite, where they are remarkably related even if the APIs are different to provide greater diversity and flexibility in terms of services offered (Figure 2-10).

---

2. CORBA is specified by Object Management Group (OMG) which is a international consortium comprising over 500 information system vendors, software developers, and users.

3. CI Labs-Component Integration Laboratories is a consortium formed in 1993 by Apple, IBM and ORACLE among others.

**Figure 2-10** Middleware functionality: The services perspective
*Middleware services are often built on other middleware services, and use other network service technologies, including management, directory and security services Each middleware service would need to have APIs for direct used by applications.*

Potentially, the use of an IIS:

(a) makes programming easier by insulating application software from complexities associated with managing of system resources. This improves portability of application programs. For example, an application can run on different network types and can be referred to in a manner which is independent of physical location (i.e. it does not matter where that application resides);

(b) provides a data communication system that allows the building of an integrated system, comprising distributed software applications. Once this system is built, the IIS enables the distributed applications to access the hardware resources in the system (which includes data repositories).

However, many existing forms of IIS offer a restricted set of integration services and tools and are to a certain extent proprietary in nature. Here the IIS can enable a flexible mapping of software applications onto the physical resources of a system. This is of major advantage with respect to enabling change, the incremental extension of a system and providing a migration path from the use of legacy software and resources towards more interoperable components. IIS offer a view of the world and integration needs held by a particular system supplier. As a result, they provide a limited 'de facto standard' interface capability where so called conformant software applications (which are strictly compatible to the IIS) will be supported. Currently this limits the potential advantage gained from proprietary forms of IIS and is particularly limiting with respect to the inclusion of legacy systems.

Furthermore, presently available forms of IIS do not treat application functions and information independently and separately. Thus changes to either function or information will inevitably affect both because of their close dependency. Indeed the task to effect the changes can be very demanding and disruptive to normal operations because careful consideration for enterprise-wide implications of such changes has to be given. This is due to the encompassing nature of the IIS towards promoting intra-organisation integration.

In recent years hardware and software developers have made a concerted effort towards facilitating a more 'open', flexible and platform-independent computing environment. Inevitably this would make it possible for much of the expectation on aspects of application software portability and distributed heterogeneous computing necessary for the IIS to be delivered more effectively and conveniently in the near future. For example, the emergence of JAVA[4], which is a platform-independent object-oriented programming language created by SUN Microsystems Inc., is a significant development. JAVA computing has been hailed as the future for network computing because of its native support for networking and ability to reduce time to market for software developers, particularly for multi- and cross-platform development. It has been strongly endorsed by many major software and hardware developers including Microsoft Corporation, Apple Computers Inc., IBM, Novell Inc., Silicon Graphics Inc., Hewlett-Packard, The Santa Cruz Operation (SCO) Inc., Hitachi Limited and Tandem Computers Inc., making the language easily accessible to developers of those platforms. Invariably, this will free the end user from the threats posed by closed proprietary systems and also the hegemony of vendor-specific operating system.

---

4. JAVA can be used to write not only applets, which are applications embedded in a World Wide Web page, but also stand-alone applications. A JAVA program is first complied into byte codes, which are then interpreted by a JAVA Virtual Machine (JAVA VM). The JAVA VM is important because it implements the applet. Thus there are JAVA VMs inside World Wide Web browsers and JAVA-enabled operating systems. Because of this concept JAVA programs are platform-neutral as they are insulated from the specific features and idiosyncrasies of various operating systems. They only interact with the JAVA VM.

## 2.3 ENTERPRISE MODELLING

It is important to model or represent manufacturing enterprises in order to describe in a formal manner the ideal situation with regard to (a) their dynamic behaviour taking into consideration the internal and external working parameters which might exert some degree of influence, (b) information requirement and flow, and (c) dependencies between functional activities in terms of workflow and processes. In addition to rationalisation, optimisation and streamlining the activities concerned, enterprise-wide transparency is also provided where uncertainties are systematically eliminated, system behaviour can be well predicated and better controlled and management of change in response to future needs can be achieved, possibly via some process of enhancement or re-engineering.

However, there is much diversity in the ideal situation in individual companies (even though they may be in the same industrial sector), in relation to the organisation structure adopted, operations carried out, manufacturing processes used, etc. Hence no two companies can be expected to be identical, each having idiosyncrasies and specific functional and information needs to realise its own business goals. Manufacturing information in particular is notoriously difficult to standardise. Notwithstanding these differences, reference models (which can serve as templates) are necessary to promote good practice and a certain degree of standardisation which can promote interoperation on an inter- or intra-company basis. Where possible these reference models should capture and describe generic properties related to 'good practice' which are widely applicable. But they will need to be open to changes to allow modification and expansion to cater for customised needs.

As reported in the literature, earlier works on reference models have been specific and functional in nature, in as much as their use has been focused relatively sharply on a specific application domain, such as integrated production planning development. For example, the use of a family of reference models was postulated by Scheer *et al.* (1991, Hars *et al.* 1992) which characterised specialised functions such as order entry, resource planning, management and scheduling. The models proposed were used successfully in facilitating requirement elicitation (this involves the identification and modelling of customisation requirements for applications) as well as enabling data re-engineering to suit specific user needs, as exemplified in Figure 2-11. Here the aim was to derive integrated production planning systems which are modular and reconfigurable in nature. Generally the reference models were not conceived with an extended application domain in mind to characterise interoperation across

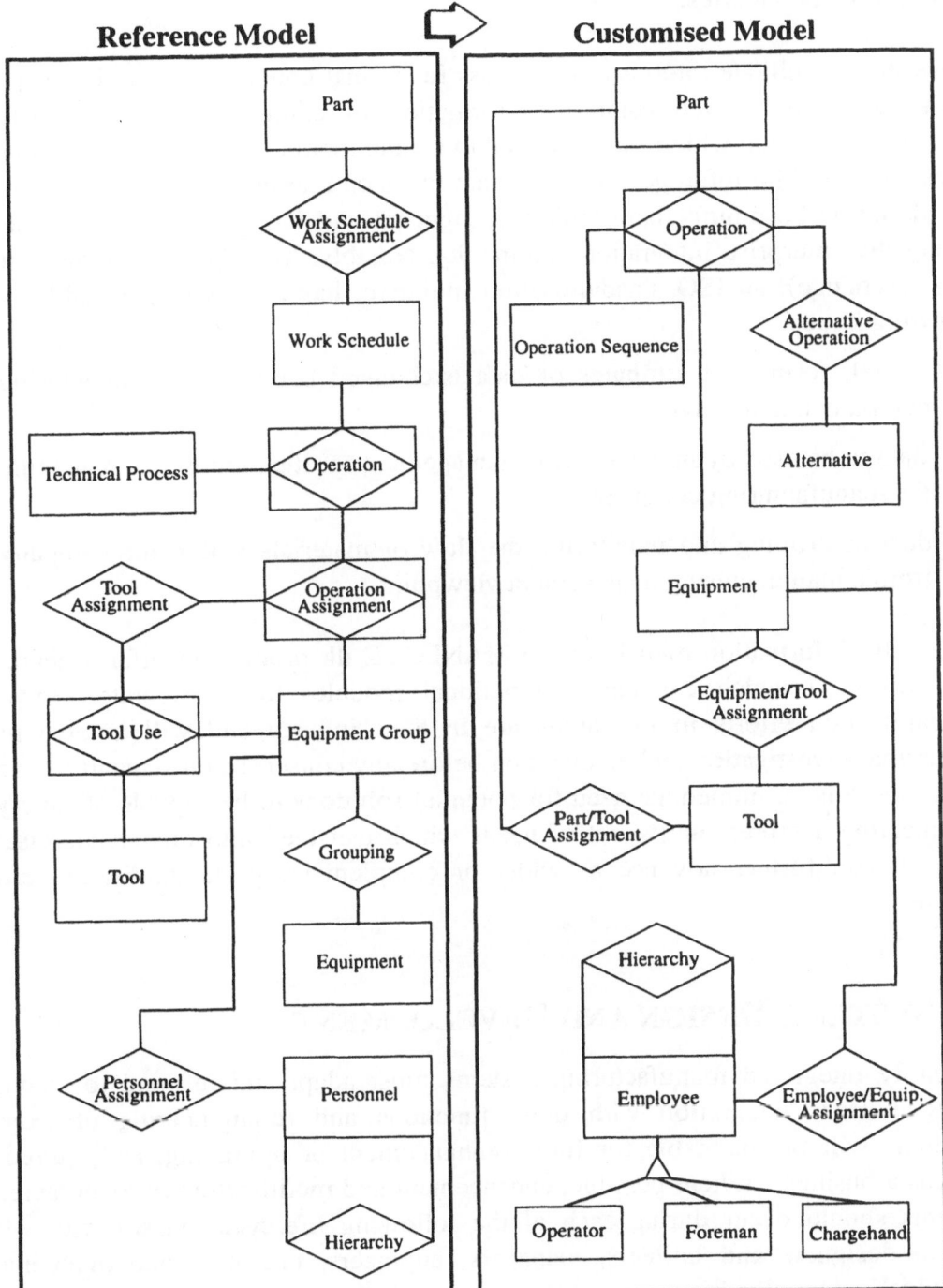

**Figure 2-11** Customising the production control area information need

conventional boundaries.

However, to facilitate interoperation across functional boundaries there is a need for reference models which comprise information of common interest to different functional areas and which can be shared to enable them to functionally interoperate. Indeed the need for reference models which focus on interoperation across functional boundaries is becoming more widely recognised and has generated much interest among the enterprise integration community. Notably, MANDATE (Manufacturing Data Exchange), an ISO standardisation initiative, has been set up to address the following issues:

- model, form, and attributes of data exchanged between a manufacturing company and its environment;

- data to be used by manufacturing management for the purposes of managing the manufacturing company;

- data controlling and monitoring the flow of materials within the company from a manufacturing management viewpoint.

Essentially information models of the MANDATE ilk promise to offer a degree of standardisation which can enable functional modules to interoperate. However, standardisation efforts in this arena are in their infancy and will involve much discussion, investigation and deliberation before any consensus on standards emerge. Hence there is an immediate need for potential solutions to be capable of satisfying and meeting a representative set of needs which can feed into this standardisation work. It can further advance as wider or complementary standardisation efforts mature.

## 2.4 SYSTEMS DESIGN AND DEVELOPMENT

In reality integrated manufacturing systems must adapt and respond to changing needs. Further integration with other functions and re-engineering of existing functions will be inevitable for future enhancement or upgrading, as required, to refocus a business. Where possible, enhancement and modification of manufacturing systems should occur during each of the following life-cycle phases with which system designers and builders, managers, engineers, operators and maintenance personnel are involved:

- **Conceptual Design** In this life-cycle phase the prime focus is deciding *what* a system should do. This can be achieved by analysing 'as-is' (present) and 'to-be' (potential) situations in order to identify means of achieving a set of improvement goals.

- **Detailed Design and Implementation** This involves specifying *how* the global requirements defined during conceptual design can be realised in terms of building the required solutions. Typically step-by-step implementation is achieved, with debugging of sub-systems carried out at each step.

- **Operation and Maintenance** This characterises the working life of the installed solution, as well as necessary adjustments and repair during the operational lifetime of the system. Generally speaking any major change will involve other upstream life-cycle phases.

Commonly each life-cycle phase is distinct in the sense that:

(a) normally different types of personnel with various perspectives and goals are responsible for each phase;

(b) various methods and tools can be employed at each stage but seldom will their use be connected through common paradigms and system models;

(c) in view of (a) and (b) 'over the fence' system engineering is a common phenomenon with major discontinuities and misunderstandings as specifications and requirements for change traverse life-cycle boundaries.

Thus current approaches to system design much reduces the opportunity to share and channel usable results and data produced in other phases. Consequently, realising life-cycle support for an integrated system progressively through its design, implementation and run-time phases is by no means trivial, especially as the complexity of a given system grows. Many of the difficulties facing system designers and builders can be attributed to the absence of a structured approach to creating systems which use common formalisms to straddle the various life-cycle phases.

## 2.4.1 Design and Modelling Methods to Support Life-cycle Phases

There are various structured design methods available which can support different life-cycle phases of integrated manufacturing systems. Typically, they structure and represent certain aspects of the system under consideration, i.e. they provide a view or views of a system, related to function, information, behaviour, etc. Commonly used structured design methods include:

- **Entity-Relationship (E-R) Modelling** This methodology was conceived to enable information modelling and can systematically convert user requirements into a set of E-R models. Subsequently the E-R models defined can be used as the underlying model for a database management system. This can help facilitate information sharing in a more structured manner where the information model may encompass information which resides in a variety of data sources.

- **Yourdon** is a widely used process-oriented methodology for designing software systems. It prescribes methods based on a set of diagrams (context, data flow, entity-relationship and state-transition) each of which illustrates a single perspective of the system. Yourdon's structured design method has been used in a variety of applications. Also Yourdon and extensions to it have been combined with other software tools and used to design integrated systems.

- **Structured System Analysis and Design Methodology (SSADM)** It has been originally conceived for software design. It is widely used in commercial applications and . The complete methodology encompasses six stages of a software project, viz: analysis, specification of requirements, selection of system options, logical data design, logical process design and physical design. To improve the input/output facilitates available to a system designer, it uses graphical modelling in the form of data flow diagrams and entity models.

- **U.S. Air Force ICAM — Integrated Computer Aided Manufacturing Definition (IDEF)** is a methodology derived from Structured Analysis and Design Technique (SADT) which has been more specifically tailored for use in manufacturing domains. Currently IDEF comprises a suite of methods which essentially can be considered under one of the following main sub-divisions (Colquhoun *et al.* 1996):

    $IDEF_0$ - This is used to produce function (or activity based) models of manufacturing systems or their sub-systems.

    $IDEF_{1X}$ - This is a data modelling methodology used to describe entities and relationships between entities.

    $IDEF_3$ - This is a dynamic modelling methodology that describes the time-variant behaviour of function blocks and information entities of a manufacturing system.

IDEF has been very widely used in a large number of industrial cases. It is used as a conceptual design modelling approach in many consultancy businesses around the world. There are on-going developments to further extend the scope of IDEF and hence its coverage of the life-cycle of manufacturing systems which includes methods for Process Description Capture, Design Rationale Capture, Implementation Architecture Modelling, Organisation Modelling and Three Schema Mapping Design.

- **Graphe à Résultats et Activités Interliés (GRAI)** is a methodology which was conceived to analyse and design production management systems. It models function, structure and behaviour with the purpose of describing the flow of information, material and decisions in systems. It includes modelling views which represent time scales in the form of planning horizons and periods. On applying the methodology to a system, a graphical model is produced which relates activities, their time frame of operation, the decisions made and the information and resources required and used.

- **Object-oriented Analysis and Design Methodologies (OOADM)** is a collection of relatively new systems design methods which are based on the object-oriented paradigm. Already in many applications they have promised to replace conventional process-oriented methodologies. Many object-oriented design methods are available that address one or more aspects of system design (either alone or combined with other methods). The main advantage of OOADM over process-oriented approaches is the closeness of the object representation to the physical system being modelled, along with its orientation towards enabling simulation.

- **Open Systems Architecture for CIM (CIM-OSA)** has been proposed by AMICE (European CIM Architecture) consortium within the ESPRIT I and ESPRIT II programs. CIM-OSA comprises a methodology and a framework which embraces the specification of an integrating infrastructure. It is suggested by certain authors that CIM-OSA 'goes far beyond previous modelling methodologies' and aims to support the design of CIM systems from their requirements definition (early stages of conceptual design) to their operation and maintenance. With CIM-OSA it is also claimed that a processable model of the CIM system can be produced as opposed to SADT-based methods which only produce static models and lack a dynamic modelling capability.

- **Business Process Re-engineering (BPR)** has in recent years come to the fore as a basis of achieving breakthroughs in a company's performance in aspects of productivity, profitability and customer service. It offers a systematic methodology for the analysis and design of work flows and processes within and between organisations to fulfil specific business requirements. There are various BPR tools available today such as ARIS, APACHE, STRIM and CADDIE. Generally they differ in characteristic and strength depending upon their emphasis for implementing process redesign. Thus it is considered that no one software package as yet covers satisfactorily all aspects of process redesign.

There have been various studies conducted which compare the capabilities of different modelling methodologies but to date no one methodology includes capabilities for modelling the function, information, dynamic and decision-making aspects of the systems. As a result, independent and separate use of a number of methods will be required if the formal modelling of systems is required on a comprehensive basis.

## 2.4.2 Entry Point for System Life-cycle Support

In summary, the application of most currently available design and modelling methodologies is primarily confined to the conceptual design phase, with a few of them extended to include limited support for the implementation phase as well. Function, information and behaviour analysis is carried out with the aim of meeting a set of previously defined requirements and goals for the system concerned. Typically the static function and information models generated using these methods will include formal definition and representation of (a) dependency relationships, and (b) configuration and composition (e.g. database schema representation and entity-attributes, resource requirements to achieve the required functions). Having obtained models of the system the effect of changes and variations can be scrutinised. Thus possible system enhancements can be identified and represented by the models.

The formal modelling of systems can provide an entry point for supporting the life-cycle of manufacturing systems where the models created (of function and information aspects) can serve as a source of knowledge during different life-cycle phases. However, in realising this potential it is necessary to develop additional life-cycle support tools coupled closely to the modelling tool. Such a software toolset should exploit the knowledge contained within the model in order to ensure compatibility and continuity between life-cycle phases, i.e. maintain consistency between the models produced and used during each life-cycle phase.

## 2.5 Means of Controlling System Behaviour

In an integrated system, functions are coupled together through sharing of common information. Formal definition of interaction between the functional components of an integrated system requires clear and accurate descriptions of (a) the flow of information between function blocks, and (b) the form and type of information, which should be made available to support and drive those functions so that they can realise their assigned tasks. Any lack of clarity very often gives rise to serious problems during system design, development, operation and changes. Hence when designing and implementing an integrated system, association between functions and information as well as dependencies among functions must be well defined and clearly captured. If this can be achieved, the relationships defined can be used to determine and govern the manner in which the system behaves, particularly during run-time.

### 2.5.1 Functions and Information Entities Association

There is a need to formally maintain an association between the functions carried out in a manufacturing system and the information entities they generate, access and manipulate. However, in manufacturing systems such associations are only formally maintained during the requirements definition and design stages of systems specification which in the software design process correspond to early phases of the development life-cycle. To address this deficiency the following proposals have been advanced by Scheer (1991) and Shunk *et al.* (1986) respectively:

- The 'Y-CIM' model proposed by Scheer provides an integral organisational view of the different subsystems of the enterprise (Figure 2-12). From such a view, the necessary links to be established between the different isolated sub-systems making the exchange of information possible can be derived. Hence the model attempts to capture and represent the information needs of associated functions.

- The Triple Diagonal concept, which is based on the use of $IDEF_0$, proposes a modification to the functional modelling approach and this includes a definition of information, control and material flows. Components of a manufacturing enterprise are classified and related to each other via a defined layered architectural relationship. By including a definition of information resource requirements as well as material, information and control flows into the $IDEF_0$ functional model, a formal association between functions and information can be defined, with the input

**Figure 2-12** 'Y-CIM' model

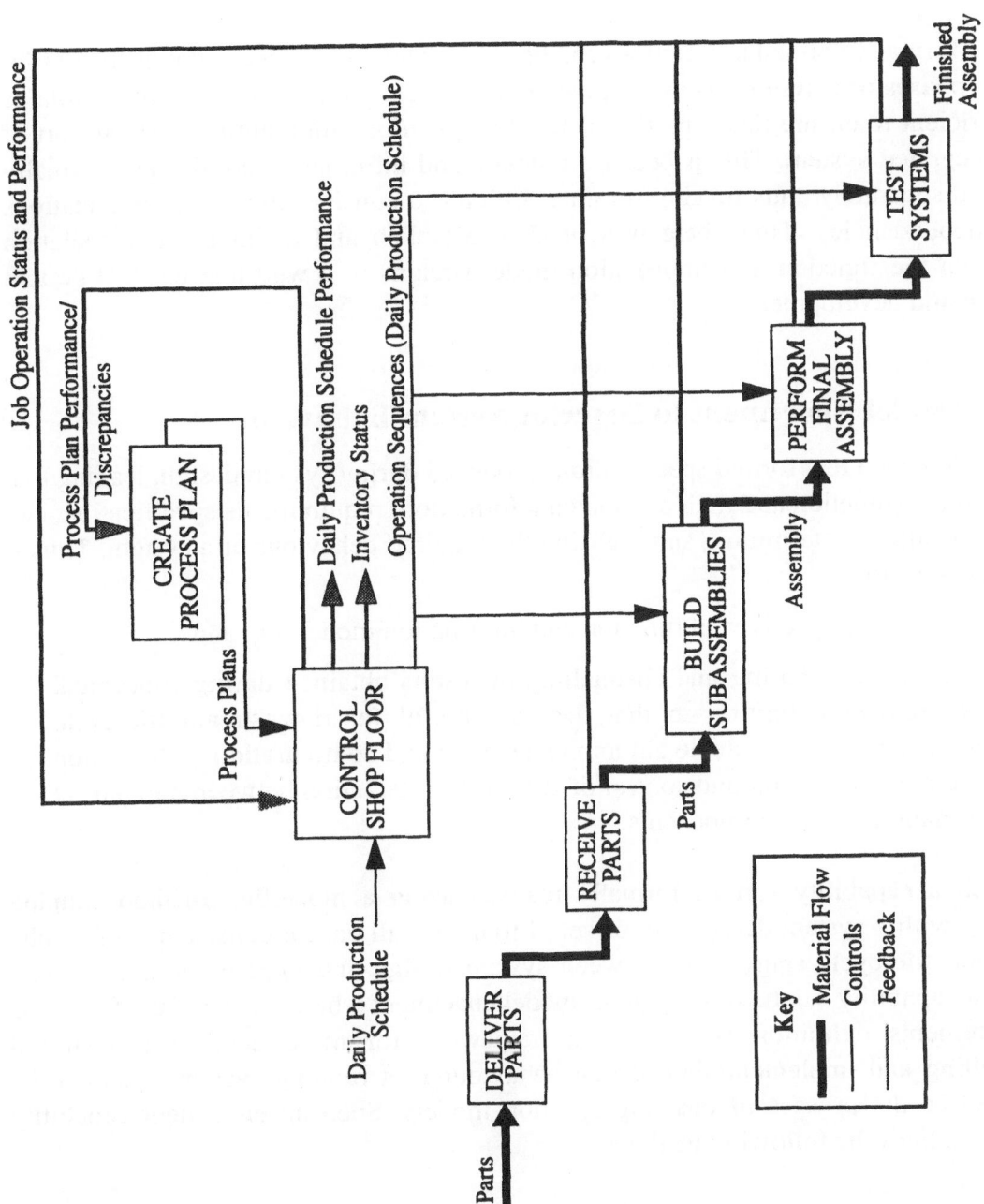

**Figure 2-13** Triple Diagonal (Material Flow/Controls/Information Integration) modelling

and output of information from associated functions being clearly identified, as illustrated in the example model of Figure 2-13.

These formally defined associations can be made available for use in downstream life-cycle phases of a system. However, use of an appropriate modelling method alone is insufficient to ensure that a particular function or information entity exists as part of an integrated system. This is because function and information entities are normally viewed separately, thus making it rather difficult to consider their close associations and dependencies. Thus there is a need to establish and maintain an association between the function and information model streams in a way that can aid system design and development.

## 2.5.2 Model Enactment to Describe System Behaviour

It is recognised that formal specifications produced during system design, leading to a conceptual (function and their associated information) requirements specification, can prove useful in determining and realising the required behaviour of a system. Thus it is necessary to:

- unify the perspectives of function and information modelling; and
- facilitate the sharing and channelling of results obtained during conceptual requirement definition so that they are useful for downstream life-cycle processes, for example, to aid implementation and configuration with relation to the co-ordination and control of functional interaction between a given set of manufacturing components.

If such a capability can be formally realised using a modelling method coupled closely with a system design tool or set of tools, it will ensure consistency of results between life-cycle stages (i.e. between system design and implementation stages). Such a capability can be referred to as model enactment where a conceptual functions requirements definition is used as a framework for more detailed behavioural modelling and implementation of that behaviour in a running system. This can be viewed as the process of enacting function models. Such an enactment capability should inherit the following benefits:

- an appropriate formal definition of interactions among functional modules based on dependencies among functions derived from higher level system descriptions and a description of data requirements to support the functions concerned;

- defined means of supporting functional interaction management based on definitions and relationships established at a higher level, thereby facilitating control system behaviour (via suitable mechanisms) by providing a description of how run-time activities can be effectively co-ordinated.

# 3  Achieving Interoperability

## 3.1 GENERAL CONSIDERATIONS

The sharing of information to create synergy among functional modules constitutes one step towards fully achieving software interoperability. In order that the benefits of software interoperability can be more fully realised, the next crucial step is to address problems of functional interaction (and the underlying issues of behavioural interaction) among functional modules. Thus the processes of co-ordinating and synchronising functional interdependency and association among functional modules (with accountability for the shared data) need to be carefully managed and controlled. In practice, this is necessary to ensure and maintain discipline and harmony, to enable co-operation among interoperating software components and to establish well defined communication channels which can collectively promote and enhance intra-organisation interaction and co-ordination of activities. The author has sought to adopt a mixture of pragmatic and formal approaches to address these issues taking into account the (a) practical problems faced by companies in trying to achieve systems integration, and (b) gaps in technology and know-how which need to be filled. As a result a novel approach has been conceived and advanced to achieve software interoperability which offers means of:

- overcoming various inherent deficiencies and constraints which would severely inhibit or complicate functional interaction; and

- tackling in a structured way integration problems associated with contemporary solutions and systems.

The aim is not to place in the foreground details of necessary functional improvements to individual manufacturing related functions but rather to focus on enabling software interoperability by providing means of building 'soft' rather than 'hard' integrated solutions.

A particular focus is on seeking an innovative methodology which can improve the reconfigurability of interoperating software applications over their life-cycle, thereby facilitating their adaptability in response to changing needs (e.g. further integration with other functions and re-engineering of existing functions in order to modify and enhance the system's functionality). Through improving the adaptability of software applications it should become possible to mitigate against early obsolescence.

Key to the methods derived will be a structuring of implementation processes based on the use of an integrating infrastructure which embodies common integration services. These common services will facilitate data management, access, manipulation and presentation, and support inter-process communication among conforming applications.

## 3.2 NUCLEUS FOR MANUFACTURING INFORMATION

Production planning is defined as encompassing the management of flows of materials and goods as well as seeking to ensure capacity utilisation based on customer orders and/or demand forecasts; this includes order entry, resource planning and management as well as scheduling functions. It is viewed as providing a comprehensive information system that offers a large pool of manufacturing and logistical data which bears varying degrees of commonality, interdependency and close association to that of other applications. For example, bill of materials (BOM)

**Figure 3-1** Production planning information as the initial nucleus

and process planning information are common information entities viewed from different perspectives by CAD/CAM and CAPP (Figure 3-1). Therefore, the choice of production planning information as the initial nucleus of a model manufacturing information is a pragmatic one. This provides a basic means of establishing data repositories to facilitate information sharing with other functional areas within a given company.

## 3.2.1 Manufacturing Continuum Consideration

Typically, the production planning methods and systems employed today are likely to have been chosen after considering particular characteristics of the manufacturing environment concerned, i.e. the type of products, production volume, demand fluctuations, manufacturing technology, markets involved, etc. Indeed there is a spectrum of discrete parts manufacturing which can be viewed as ranging from 'Project Manufacturing' to 'Repetitive Manufacturing' (Figure 3-2). Not surprisingly therefore there is considerable diversity and lack of uniformity among production planning methods and systems.

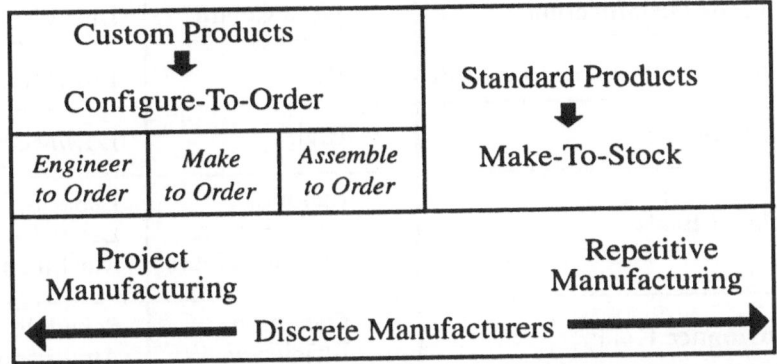

**Figure 3-2** Continuum of discrete part manufacturing environments

Clearly, the environment required for each class of manufacture (be it 'engineer-to-order', 'make-to-stock', etc.) will require its own distinct set of operating characteristics, as illustrated in Table 3-1.

Hence suitable manufacturing methods have to be adopted to satisfy and address the specific needs of each class of manufacture. They govern the formulation and determine the nature and extent of functionality, particularly with regard to production planning and control.

| | Manufacturing Environment | |
| --- | --- | --- |
| | *Engineer-To-Order* | *Make-To-Stock* |
| **Production Complexity** | High | Low/Medium |
| **Capacity/Material Driven** | Material | Capacity/Material |
| **WIP Value** | High | Low/Medium |
| **Push/Pull** | Pull | Push |
| **Schedule vs Orders** | Orders | Schedule |
| **Forecast Stability** | Low | Medium/High |
| **Direct Issue vs Backflush** | Direct | Backflush |
| **Shop Floor Organisation** | Work Centre | Line/Cell |
| **Manufacturing Operations Staff** | High | Low/Medium |
| **Labor Content** | High | Low/Medium |
| **Overhead Basis** | Labor $, Labor Hours, Machine Hours | Labor $, Labor Hours, Machine Hours |
| **Performance Goals** | Operation Efficiency | Schedule Attainment |

**Table 3-1** Characteristics of discrete manufacturers

Common manufacturing methods adopted include Manufacturing Resource Planning (MRP II), Just-in-time (JIT), Optimised Production Technology (OPT) or possibly a hybrid of them where:

• **MRP II** operates on forecasts, time-phased resource planning and fixed lead times. It is suitable for discrete part manufacture of standard products for the 'make-to-stock' environment. However, it does not include functionality to support short term scheduling of orders under the constraints of the real availability of resources. This can make the use of MRP II inflexible, capacity insensitive and not responsive enough to changes and short term demands imposed by, for example the 'configure to order' environment, which is characterised by small batch quantities and large product varieties.

- **OPT** is based on the insight that the flow of materials and goods, consequently affecting the performance of the manufacturing system, is determined by properties of its bottlenecks, such as limited capacity, demand and availability of raw materials.

- **JIT** methods are not only used to reduce inventories but also for continual improvement of the production process. This approach requires a reorganisation of the logistic chain to provide sufficient flexibility and reliability to closely match resources and capability to customer demand.

The strength of MRP II lies in its mid- to long-term global planning whereas the strengths of JIT and OPT are in the short term execution of planned needs. Each has proved to be successful in certain production environments and each has demonstrated disadvantages under certain conditions. It is beyond the scope of this book to discuss further these paradigms and as each of them is sufficiently complex, no attempt will be made here to elaborate upon the various philosophies which they embody. The literature available on these paradigms is well supplied, thus the reader is advised to refer to them to gain a better understanding.

Bearing in mind the following realities, there is a need to scrutinise the (information and functional) requirements of the various manufacturing methods so as to identify amongst them (a) essential information of prime concern, and (b) the existence of common threads of functionality:

- In practice companies do not necessarily stop at the boundaries of one manufacturing method but cross and mix ideas to extract what makes sense for that particular company

- Any choice of manufacturing method depends upon various drivers (e.g. the nature of the management style, organisation, information technology and manufacturing technology), each of which will inevitably change and evolve, as illustrated in Figure 3-3.

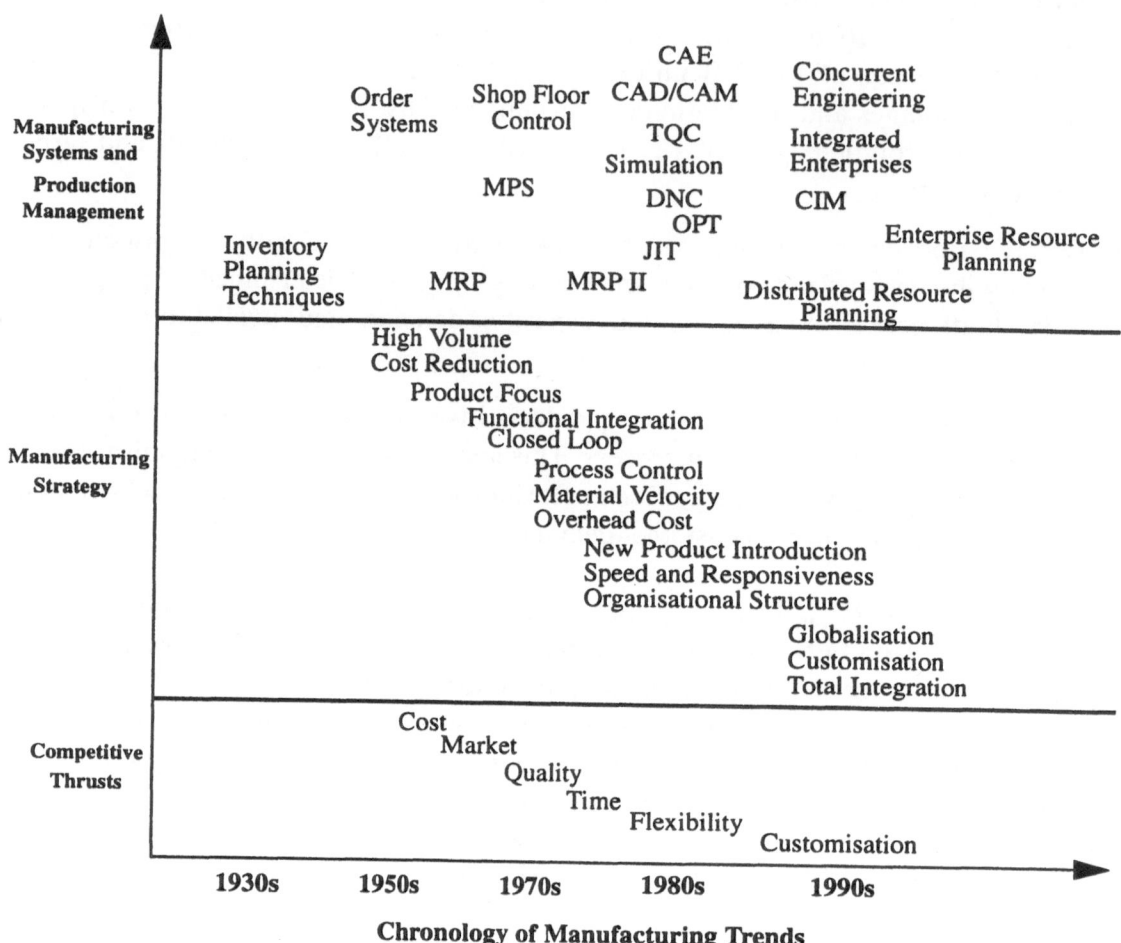

**Figure 3-3** Evolution of production management and manufacturing systems

Quite importantly, such generalisation would alleviate any bias towards specific manufacturing methods, thereby effectively breaking free from the restrictive bounds of the manufacturing methods adopted. In addition, commonality in information requirement and functions would indeed help overcome the considerable diversity among production planning methods and systems, currently existing across the manufacturing continuum, which serve to inhibit the achievement of interoperable solutions.

## 3.2.2 Manufacturing Methods and Information Requirement

It is interesting to note that the information requirements of adopted manufacturing methods, which include MRP II, OPT and JIT, demonstrate many similarities, differing mainly in their emphasis and degree of focus on the activities concerned. This is illustrated as follows:

- OPT is a computerised scheduling system which employs a standard MRP style database of BOM, resources, routes (including data on setup and operation times), inventory (raw material, work-in-progress and finished product) and demand (specified by due date and quantity required)

- JIT, with its cellular manufacturing approach and the use of a 'demand pull' concept to control the production and movement of parts through the production process, requires information on production schedules (specified by start and finish dates on a daily basis), BOM, inventory (specified by lead-time for raw material and components delivery), routes (includes data on cell output, setup, process and cycle times) and capacity (specified by available working capacity for loading parts for manufacture).

The manufacturing method adopted in a given company will significantly influence the functional requirements of the manufacturing related applications. Similarly, characteristic properties of each manufacturing environment will directly influence these needs. This gives rise to considerable diversity among the functional properties of different production planning methods and systems which currently exist to support the manufacturing continuum. Hence when looking for similarities, it is more appropriate to consider and focus on similarities with regard to their information requirement.

## 3.3 INFORMATION MODELS

Ideally it is necessary to make available a pool of most of the information entities that are commonly shared between software applications. They should capture and embody knowledge concerning 'good practice' or 'good solutions' and be generic in nature so as to be applied to enterprises within a particular industrial domain. The aim is not only to create synergy among the functional modules concerned but also to guide and aid in building and improving integrated manufacturing systems, thereby reducing the time involved and improving the quality of the resulting system.

Hopkinson (Evans *et al.* 1993), in his analysis of user needs relating to information standards, strongly stressed the point that:

> *What matters most of all for the user is the information the system holds; the way in which it is held and accessed, and what can be done with it, are also important but the means are of no value if the information itself is not what is needed.*

In order to achieve this an information model is required which identifies real world objects, their key attributes and inter-relationships. Indeed information models which constitute essential information of common interest (particularly among production planning, SFCM, process planning and finite capacity scheduling) have been conceived and defined by the author. The information models represent information entities and their inter-relationships and they can be used as a generalised resource in many discrete parts manufacturing environments.

The models defined represent shared information which are 'items of knowledge' essential for part manufacture. These have global interest to the functional modules concerned. For example, a fundamental requirement during many product design processes is to ensure the manufacturability, ease of assembly and testability of the product. Therefore, it is necessary for the designer to have access to accurate information about the manufacturing process, with due consideration to constraints such as the availability and type of resources (e.g. material, toolings and fixtures) and orders which exist for the product. For example, information stored in conformance with the models will be useful in matching product specifications and requirements to the capabilities available in the enterprise. This fundamental requirement is not usually provided for, which is a major cause of a fairly large proportion of engineering changes and discrepancies in manufacturing enterprises today.

## 3.3.1 Specification

With respect to the discrete parts manufacturing domain, the author carried out a detailed analysis of production planning function. This study considered common information and functional inter-dependencies in typical product design, process planning, finite capacity scheduling, and SFCM processes.

The study involved a review of the literature and current practice with regard to the techniques used to accomplish integration of production planning, CAD/CAM, CAPP and SFCM systems. Some of the findings of this work are summarised in Table 3-2.

| Relevant References | Integrated Systems | Commonality of Information | Functional Purpose and Dependency |
|---|---|---|---|
| Halevi and Weil 1992<br>CIM Strategies 1991<br>Scheer 1991<br>Singh 1991<br>Harhalakis *et al.* 1990<br>Lang-Lendroff *et al.* 1989<br>Scheer 1989<br>Schnur 1987<br>Bohse and Harhalakis 1987<br>Ssemulaka 1987<br>Logan 1986<br>Saxe 1985 | Integration of Production Planning with<br>- CAD/CAM<br>- CAPP<br>- Shop Floor Control Systems | Bill of Materials<br>Inventory and Resources<br>Process Plans/Routes<br>Schedules<br>Shop Floor Status Feedback | 1. Product hierarchical decomposition into sub-components to aid<br>a) resource planning;<br>b) sub-components manufacture; and<br>c) standard components procurement.<br>2. Material requirement planning and allocation for part manufacture (e.g. raw material, toolings, fixtures, manufacturing facilities, etc.).<br>3. Sequencing of manufacturing activities for part manufacture. |
| Zäpfel and Missbauer 1993<br>Lee 1993<br>Muhlemann *et al.* 1991<br>Scheer 1991<br>Ptak 1991<br>Waterlow and Monniott 1986<br>Luscombe 1991 | Integrated Production Planning with Modularised functions | | Material requirement planning<br>Resource allocation and scheduling<br>Order entry<br>Routing<br>Capacity planning<br>Inventory management<br>Shop floor status monitoring and acquisition |

**Table 3-2** Summary of literature review

In addition, the study drew upon insights gained from the author's previous industrial experience. Among successful implementation projects undertaken by the author, a particularly valuable source of reference input was derived when developing a CIM model for the Singapore Economic Development Board (between 1989 and 1992). This model is still used today to provide a technological showcase for Singapore's precision machining industry for discrete part manufacture. A general overview of the functional properties and the information entities of this CIM model is illustrated in Figure 3-4.

The CIM model was developed to support state-of-the-art shop floor flexible machining and assembly cells. The two cells are linked via an automated material handling system which consist of an automated storage and retrieval system (AS/RS) and automated guided vehicles for part transfer. A heterarchical control architecture was adopted to facilitate 'pull' manufacturing on the shop floor (with services and resources allocated based on need and availability). The strategy adopted was to enable distributed SFCM with a capability to achieve (a) localised control and autonomy of the various shop floor devices, and (b) peer-to-peer communication in an event-driven manner for co-ordinated material flow. In addition, various enabling and emerging computer-aided technologies relating to business, engineering and production were successfully incorporated into the demonstration system to facilitate domain-specific application software interoperability and information flow across functional boundaries, which includes production planning and control, process planning, CAD/CAM, tool management, distributed numerical control, shop floor scheduling, process simulation and SFCM. Information shared among the functional modules of the system and their information dependencies is illustrated in Figure 3-5.

This large scale ambitious project involved close collaboration with various hardware and software vendors and manufacturers as well as industrial end users. It captured many of their key needs and led to an accepted and intrinsic representation of activities to support part manufacture, i.e. universally useful in understanding important aspects of interoperability in a manufacturing enterprise.

The system has served to:

- globally specify information inputs and outputs and those shared among manufacturing related functional modules;
- identify close relationships and dependencies among typical functional components;

**Figure 3-4** Overview of the functional and information network within the CIM model

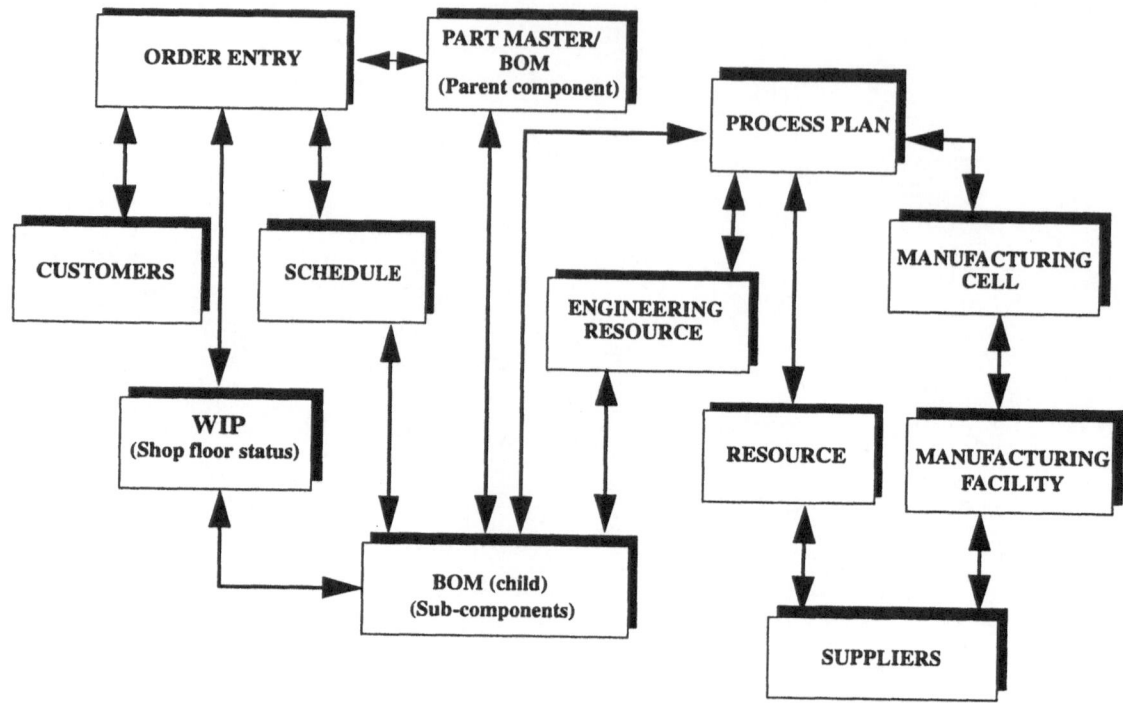

**Figure 3-5** Overview of information flow and data dependency

- define data mapping requirements based on functional relationships;
- ascertain which computer-aided tools are available and which must be tailored or developed to enable and enhance integration processes; and
- identify a functional model which is representative of precision machining m enterprises.

As a result of close collaboration with industrial end users, the CIM model project was successful in capturing and reflecting their practical requirements in terms of shared information requirements to facilitate interoperation among the functional mofiles concerned. This consideration (of the end users viewpoint) is very important in helping to validate the findings derived from the literature and current practice reviews — this is important in order to achieve a pragmatic and industrially acceptable solution.

The author's study of generic production planning functions was also based upon an examination of generic features and common attributes among representative commercially available computer-aided production management (CAPM) packages. Those selected for detailed study are listed in Table 3-3. The choice of CAPM products was made with a view to ensuring that collectively (a) they encapsulate the generic working knowledge of a number of vendors which itself reflects the perceived needs of many manufacturing user organisations, and (b) a broad basis of technical support was provided and implicitly covered the information and functional requirements of a wide variety of product types.

| CAPM Sub-systems | MANMAN | EZ_MRP | ELMS | CIMM | COMET | MFG/PRO | Fourth Shift | MCC | AVALON | Available Modules |
|---|---|---|---|---|---|---|---|---|---|---|
| Planning | ✓ | ✓ | ✓ | ✓ | ✓ | ✓ | ✓ | ✓ | ✓ | Capacity Requirement Planning |
| | ✓ | | | ✓ | ✓ | ✓ | ✓ | | ✓ | Master Production Schedule |
| | ✓ | ✓ | ✓ | ✓ | ✓ | ✓ | ✓ | ✓ | ✓ | BOM |
| | ✓ | ✓ | | ✓ | ✓ | ✓ | ✓ | ✓ | ✓ | Manufacturing Order Management |
| | ✓ | ✓ | ✓ | ✓ | ✓ | ✓ | ✓ | ✓ | ✓ | Routing |
| Inventory Management | ✓ | ✓ | ✓ | ✓ | ✓ | ✓ | ✓ | ✓ | ✓ | Material Requirement Planning |
| | ✓ | ✓ | ✓ | ✓ | ✓ | ✓ | ✓ | ✓ | ✓ | Inventory Control |
| Scheduling | ✓ | ✓ | | ✓ | ✓ | ✓ | ✓ | ✓ | ✓ | Scheduling |
| Shop Floor | ✓ | ✓ | | ✓ | ✓ | ✓ | ✓ | | ✓ | Shop Floor Control |
| Manufacturing Support Services | ✓ | ✓ | ✓ | ✓ | ✓ | ✓ | ✓ | ✓ | ✓ | Sales Order Processing |
| | ✓ | ✓ | ✓ | ✓ | ✓ | ✓ | ✓ | ✓ | ✓ | Purchasing |
| | ✓ | ✓ | | ✓ | ✓ | ✓ | ✓ | | ✓ | Financials |
| | ✓ | ✓ | ✓ | ✓ | ✓ | ✓ | ✓ | ✓ | ✓ | Management Reporting |
| | ✓ | | ✓ | ✓ | ✓ | ✓ | ✓ | | ✓ | Costing |

**Table 3-3** CAPM packages examined

However, inherent difficulties were experienced in establishing a common basis for formally evaluating and analysing these candidate systems because of major differences in their underlying philosophies and their implicit understanding of the activities within a factory and how they should be controlled. This is further complicated by the proliferation of names used to refer to essentially similar and basic functional capabilities which collectively enable production management. Also in different systems different functions can be included under the same name. Thus the

*The CIM Debacle*

CAPM systems analysed were compared and classified with reference to the following sub-systems in order to gain a more structured and uniform understanding of the functions offered:

- Planning
- Inventory management
- Scheduling
- Shop floor
- Manufacturing support services.

Following this analysis common classes of functional module and information entity were identified, as listed in Table 3-4. As a result information models were identified and defined to satisfy generic requirements, as listed in Table 3-5.

|  | **Production Planning** |
|---|---|
| **Product Design** | * Inventory / Part Master Records<br>* Bill of Materials |
| **Process Planning** | * Process Plans / Routes<br>* Manufacturing Facility Records |
| **Finite Capacity Scheduler** | * Manufacturing Orders<br>* Bill of Materials<br>* Work Centre Capacities |
| **Shop Floor Control and Monitoring** | * Scheduled Manufacturing Orders<br>* Shop Floor Production and Status Feedback |

**Table 3-4** Commonality of information

Further details on the information entities and attributes represented in the information models are included in Appendix I. Examples of associations between these information models and specific information models which form the basis of the commercially available Manufacturing Machine Code (MCC) and EMM Lane Manufacturing System (ELMS) computer-aided production management application

| | |
|---|---|
| **Manufacturing Facility** | Information on manufacturing support facilities with data on manufacturing capabilities and specification. |
| **Part Master/ BOM** | Product structure according to its sub-components relationship. |
| **Resource** | Inventory record and status for raw materials, fixtures and tools inclusive of labour and facilities, i.e. work centres and processes. |
| **Process Plan** | Routing for part manufacture. It includes the sequence of operation for planned and alternative processes and manufacturing resource requirement. |
| **Order Entry** | Order registration for order type, quantity, batch size and due date. |
| **Schedule** | Production timetable where manufacturing orders are scheduled according to order commitment and availability of resources. |
| **WIP** | Shop floor status feedback, actual to planned comparison, and work centre utilisation rate. |
| **Engineering Resource** | Includes resources related to or assigned for part manufacture, e.g. engineering drawings and NC programs. |
| **Manufacturing Cell** | Grouping of manufacturing stations for manufacture of a family of products. |

**Table 3-5** Information Models

software are included in Appendix II. MCC and ELMS are for the control of production, especially of material movement where they focus on the shop floor, taking into consideration the availability of resources (labour and machines), and the production process (what operations must be performed, their order, how they are linked together, their standard times, etc.). MCC and ELMS were primarily chosen because their respective developers were able to provide transparency to the internal schema of their relational database so as to help better understand its proprietary data representation and mapping.

A case study (discussed in detail in Section 4.2) has been carried out in collaboration with the University of Bradford Management Centre, UK, to validate the applicability of the generic information models. In this case study, the ELMS CAPM software package has been re-engineered with reference to the generic reference models to facilitate its ease of further development and to enhance its existing functionality.

System design and modelling tools have been used to formally represent and structure the function and information models defined in Tables 3-4 and 3-5. $IDEF_0$, which is well suited for activity-based modelling, has been used to define (a) inter-dependencies and close relationships among common classes of functional modules identified in terms of their associated inputs and outputs, and (b) information flow as seen from the viewpoint of a systems designer. In addition, $IDEF_{1x}$ was used to represent information requirement in terms of entity-attribute relationships for the information models concerned. The choice and application of these software tools is further discussed in the subsequent chapters. An overview of the IDEF methodology is provided in Appendix III.

It is important to stress at this juncture that the generic information models identified and formally defined provide an important cornerstone to the author's overall approach to enabling software interoperability. Furthermore, the author is confident that the components of those models and their inter-relationships are appropriate, certainly with respect to their use for the forms of functional modules investigated here. However, it is not argued that the models are sufficiently definitive or complete to form the basis of a standard model, but they are sufficiently definitive and complete to contribute towards an important advance in creating more open and configurable forms of integrated manufacturing systems and as such can provide a reference model of good practice which can be refined and enhanced, possibly until it reaches the status of a standard. Also later in the book it will become clearer that a second cornerstone of the author's approach is the use of model enactment to guide integrated system life-cycle processes. Indeed through the use of formal models and a set of tools that can manipulate and transform those models, improved opportunities exist to refine and enhance a reference model until it becomes more widely accepted and used.

## 3.3.2 Characteristics of the Information Model

The information models identified can be described as reference models because they incorporate essential data (i.e. data of prime concern) that will very commonly be used in the discrete parts manufacturing domain. They can be used to guide the specification and development of more specific information models, thus serving as a foundation upon which the rest of the enterprise data can be built. Together, these information models can provide the enterprise with a single coherent view of engineering, production and management information which will be in common usage throughout the product life-cycle. Data will need to be stored with reference to these generic models and in a practical system will be stored in distributed data repositories to enable common access and usage by the functional components.

It is recognised that information is notoriously difficult to standardise. Therefore, as outlined earlier the choice of information models is not meant to be fixed and exhaustive in nature, rather the models have been chosen to serve as generic reference models which are open to changes and can be modified and expanded when necessary. Bearing these restrictions in mind the information models can possess the following characteristics:

- **Wide Applicability** They conform to the requirements of many potential users and are not structured or geared towards a particular enterprise but rather for a set of enterprises. In this study the reference models where chosen to support the precision machining industry but they could well have an essential form which can constitute the basis of reference models for other industrial sectors.

- **Flexibility** They are adaptable and can be customised to specific needs of a user. The flexibility is attained as a result of their formalism in a computer readable form offering opportunities to manipulate their underlying data structures.

Also collectively these information models can serve as partial models which can be further expanded or coupled with specific information models that contain information which is unique to the manufacturing environment concerned. Certainly before the models advanced here could form the basis of any standard it would be necessary to consider the extent to which a reference model should aim to be complete, as the inclusion of entities seldom used will inevitably lead to some overheads in terms of required data storage and processing capabilities.

# 3.4 NEED FOR INTERCONNECTION AND INTEROPERATION

Functional software components are required to be interconnected in an effective manner before their interoperation can be enabled. A low level data inter-communication and information transfer facility (which facilitates data transfer over a digital link) is an important prerequisite to efficiently interconnect the software components. However, generally such a digital data transfer capability needs to be built upon in order to facilitate interoperation in a controlled and deterministic manner.

This enhancement can be realised through the use of an IIS whose purpose is to provide structured access to information services in a way which simplifies interconnection between the component elements of software systems. Ideally an IIS should comprise the following two levels of integration mechanisms and tools:

- **Low Level** This can encompass a number of general purpose means of accomplishing the integrated operation of computer software processes (or software applications). In manufacturing enterprises (as in many other computational systems) software applications will be embedded in equipment and computer systems. Hence the low level mechanisms need to resolve differences arising from heterogeneity in computer processing hardware, software, operating systems, networks, human interface systems and data sources supported. They will be required to support appropriate low level protocol between interacting software applications as well as to resolve differences in representing and storing information.

  It would probably be worthwhile to note that CORBA is rapidly emerging as the industry standard for systems integration at this level. Briefly, CORBA is object-oriented in nature and encapsulates easy to use low level distributed computing and integration mechanisms by which objects transparently make requests and receive responses. The CORBA ORB provides the possibility of interoperability between applications built in different languages, running on different machines in heterogeneous distributed environments. Some of the key benefits to CORBA-based systems integration include faster system delivery, enhanced software reuse, increased system adaptability, enhanced system interoperability and reduced maintenance costs.

- **High Level** This includes high level integration mechanisms and system management tools to more directly facilitate the interoperation of the software components (which is one of the primary issues being discussed in this book). They embody domain knowledge (relating more specifically to manufacturing systems integration) to enable the integrated operation of manufacturing applications and their internal threads of application functionality. However, the high level mechanisms need to be built upon their low level counterparts.

Figure 3-6 is a generalisation defined by Weston (1993) which distinguishes low level IIS mechanisms and tools from their high level counterparts.

**Figure 3-6** Components of an integrating infrastructure

A key aspect of methodology in this book is to facilitate software interoperability as the structuring of integrated system implementation processes based on the use of an IIS.

## 3.4.1 Requirements of Integrating Infrastructure

In 1990, research from Manufacturing Systems Integration Research Institute (MSI), based at Loughborough University of Technology, UK, led to the development of the CIM-BIOSYS (CIM-Building Integrated Open Systems) IIS (Weston *et al.* 1988, Clements *et al.* 1993) which, as depicted in Figure 3-7, achieves a unification of general purpose computational integration mechanisms and tools. The author through his close research collaboration with MSI has since contributed to this research study (SI Group 1994, Singh 1994) carried out on the general requirements of integrating infrastructure for manufacturing systems integration.

CIM-BIOSYS IIS has been used to create a variety of 'proof-of-concept' and 'live industrial' integrated systems. This IIS provides a means for structuring, decomposing and simplifying solutions and supporting their run-time execution and change. Of particular importance has been an implicit ability to build and modify systems (including systems of very wide scope) on an incremental basis. The use of the IIS has demonstrated significant savings in the cost and time involved in manufacturing integration projects.

However, CIM-BIOSYS IIS only included low level integration mechanisms and tools, as depicted in Figure 3-6. It embodies common integration services to facilitate data management, access, manipulation and presentation, and support inter-process communication among functional components. In isolation, the CIM-BIOSYS IIS can only facilitate bottom-up system build; hence by building high level integration mechanisms upon its low level counterparts, top-down system design and construction can be more readily facilitated.

The CIM-BIOSYS IIS is highlighted here merely to serve as a reference with the main intention of providing a tangible insight towards better understanding the general features and functionality expected of the IIS. Generally, CIM-BIOSYS IIS compares favourably with some commercially available solutions where it offers a range of development tools and APIs for programmers and third-party developers to:

- help simply and structure interconnection between software applications;
- facilitate consistent user and device interface; and
- enable access to common integration services for information access and management as well as inter-process communications.

**Figure 3-7** CIM-BIOSYS integrating infrastructure:
(a) functional view; and (b) operational view of CIM-BIOSYS IIS

The IIS offers important advantages over contemporary turnkey and custom built integration methods, in that inherently it:

- **Deals with Complexity** Applications only need knowledge of how to access the platform, rather than how to access '$n$-1' other applications within the integrated system. This results in a vitally important means of coping with increased complexity as the system complexity will grow in proportion to the number of applications rather than the square law fashion found using contemporary approaches.

- **Copes with Change** It removes integration knowledge from interacting applications concerning the actual structural relationships, interaction mechanisms, information structures, data formats and communication protocol; the integrating infrastructure deals with such issues. This knowledge is placed in the form of configuration data which can be used in a systematic way to enable and support change.

- **Promotes Standardisation** This is achieved by specifying a consistent interface between the services of the integrating infrastructure and the applications which use them. Also the integrating infrastructure is itself decomposed into more manageable sub-systems which can be standardised or built on existing standard mechanisms and services. Thus applications can be treated essentially as open applications and as such they themselves can become standard building blocks of systems.

Application components (i.e. software packages embedded with the controller of machines, or human interface software elements) which are not inherently compatible with the IIS architecture are considered to be non-conformant alien applications. 'Drivers' and 'alien application shells' comprise software which respectively encapsulates resources (such as database and datafile systems) and legacy application components and provides them with sufficient capability that they can use the integration services of the IIS. This would enable legacy systems to be included together with those conformant software application, thus helping to safeguard a user's previous investment in computer systems. However, typically an encapsulated legacy component will not be as readily configurable nor as completely integrated into systems as will be counterpart IIS conformant software components, built in a way which can directly utilise the configuration and integration services of the IIS.

# 3.5 AN OVERVIEW OF THE METHODOLOGY DERIVED

A meta-level overview of the methodology adopted and developed by the author to facilitate integration and enable interoperation among manufacturing related functional modules is depicted in Figure 3-8. This methodology comprises five inter-related and consistent sub-methods which collectively structure and support key aspects of integrated manufacturing system design, build, operation and change management. One of the underlying concepts adopted in the methodology is that (as far as possible) application functionality will be decoupled from capabilities included to access and update information repositories to enable the information to be treated independently from the functional capabilities realised by software applications. This not only enables easier access to information but also decouples changes to application processes from those associated information systems. The purpose of the sub-methods is outlined as follows:

## 1. System Specification

A set of high level modelling methods, based on a set of generic reference models, is used to facilitate system design. The output from this specification stage will be separate activity-based and entity-attribute models of 'functions' and 'information entities' and their inter-relationships.

## 2. Means of Enacting Function Models

A set of build tools are used to create executable descriptions of the system behaviour where the descriptions are consistent with the function models generated by (1). The output descriptions can be used to control the way in which functional components interact during system run-time.

## 3. Means of Enacting Information Models

A second set of build tools are used to create and populate information models in a form which structures and enables control of information shared among functional modules during system operation. The output descriptions are consistent with the information models created in (1) and with the outputs generated from (2).

## 4. Use of an Integrating Infrastructure

This facilitates run-time operation in a flexible data- and event-driven manner, mapping distributed software solutions onto physical resources. Here the IIS is

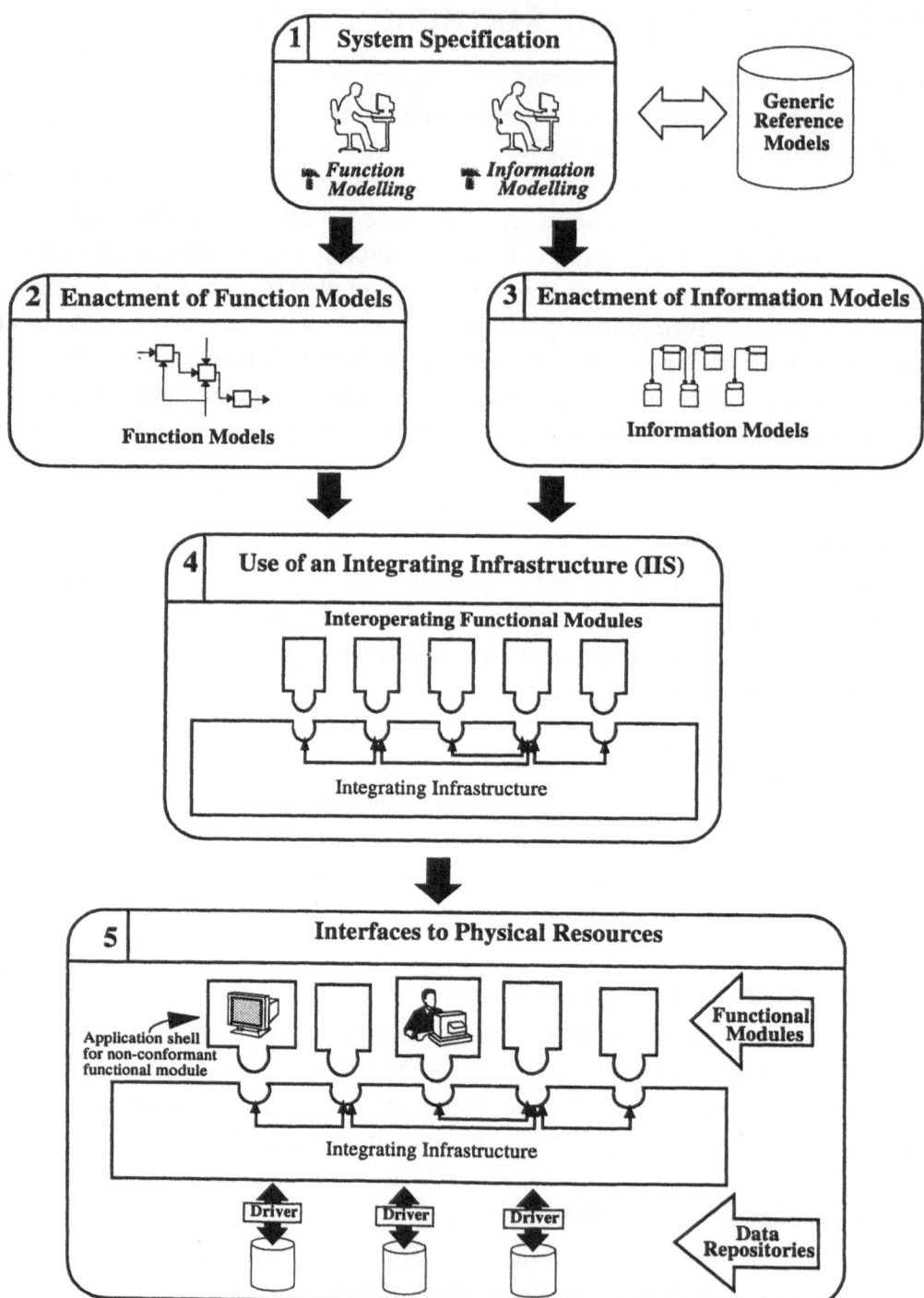

**Figure 3-8** Overview of methodology

charged with resolving differences in the physical system relating to heterogeneity, distribution and data fragmentation. Indeed the use of the IIS is the key to the methodology. As the IIS assumes responsibility for maintaining knowledge of integration details (e.g. networks used, the hardware and operating systems on which a functional component is run, the location of an information fragment, etc.), the functional components themselves need only have knowledge of how to use the IIS (i.e. not of each other). Essentially this leads to a linear relationship between system scope (in terms of the number $n$ of functional components) and complexity as opposed to the square law relationship inherent in 'pair-wise' integration methods.

## 5. Interfaces to Physical Resources

An essential element of this methodology is the ability to flexibly map software applications onto system resources, i.e. database and computer hardware. 'Drivers' and 'alien application shells' represent the software tools created to bring resources and components (which include proprietary software packages, database and datafile systems) to a level of conformance which enables interoperation over an IIS. This can provide a migration path towards more 'open' components at a later stage.

# 4 Information Architecture

## 4.1 MODELS FOR EXTENDED APPLICATION DOMAIN

The reference model approach adopted here has an extended application domain in mind which crosses conventional product boundaries and common organisational boundaries found in many manufacturing enterprises. Thus the reference models specified (in Section 3.3.1) comprise information shared by several functional areas and effectively serve as a precursor to enable interoperation. In comparison, many of the other approaches advocated and validated have been specific in nature, in as much that they have focused relatively sharply on an application domain, such as integrated production planning development.

The perspective and inputs gained from on-going standardisation initiatives, such as MANDATE to model manufacturing information (refer to Section 2.3), will undoubtedly help to further advance the generic reference models conceived by the author. These reference models aim to provide an effective solution capable of satisfying current needs.

## 4.2 CASE STUDY: APPLICATION OF REFERENCE MODELS

ELMS is a commercially available CAPM software which was conceptualised at the University of Bradford Management Centre (UBMC), UK. It was developed as a set of integrated software templates which comprise the following 'core' or principal production management functions:

- Production planning (which includes material requirement planning);
- Production progressing;
- Materials management;
- Costing.

The ELMS software application has been developed based on the use of a proprietary relational database, namely DP4, and a fourth generation language, namely Datafit, to aid data manipulation, representation and access. It consists of a group of executable programs which operate on the database in order to realise the functionality of the 'cores'.

There was, however, a need to enhance the basic functionality of ELMS to provide a resource scheduling capability. The work entails the development and incorporation of the required scheduling capability where its information needs could be satisfied and derived from the existing underlying database. Due to the proprietary nature and lack of understanding of the database structure (in terms of the information entities represented and their dependency and inter-relationship defined within the database), the task of identifying the relevant information necessary to support resource scheduling has proven to be very difficult and demanding. This problem is typical for such 'as is' software systems, a property which can severely inhibit their future development.

As an initial step towards facilitating ease of further development to incorporate resource scheduling capability in ELMS (and indeed future functional enhancements as required), work has been carried out at UBMC to restructure the ELMS database with reference to the generic information models proposed by the author in Section 3.3.1. As illustrated in Figure 4-1, this research has involved the following activities:

(a)  identification of specific information entities from the ELMS proprietary database which correspond closely to those represented in the generic information models (refer to Figure 4-2 for illustration);

(b)  establishing a mapping between those closely associated information entities (refer to Figure 4-3 for illustration);

(c)  populating with data the mapped information entities (contained in the generic information models) with relevant physical data from the existing ELMS database.

Based on this restructuring, the information entities required for resource scheduling (which are stored with reference to the generic information models) are listed in Table 4-1.

This case study clearly demonstrates the ability of the generic reference models (identified by the author) to:

**Figure 4-1** Restructure ELMS database with reference to generic information models

**Table 4-1** Information requirement for resource scheduling in ELMS

**Figure 4-2** Grouping of data represented in ELMS database with reference to information models

**Figure 4-3** Mapping between information entities

- provide clarity in the database schema where information entities and their attributes are clearly defined so that they may be well understood within the enterprise;

- increase the flexibility of the database to enable information required by the functions concerned to be easily and quickly available (with flexible association between information entities); and

- transform from a proprietary database schema to a more widely applicable and formally structured schema for which future change can be more readily supported.

One of the major benefits of the approach adopted is an enhancement of ELMS from a stand-alone 'as is' CAPM software application to one which now has the potential to more readily interoperate with various other manufacturing related applications. This is made possible because the information stored with reference to the generic information models in the underlying database constitutes information of common interest which is typically shared between various other functional modules concerned (e.g. product design, SFCM and process planning systems). From the results of this case study the author is confident that the information model and the method used to transform between proprietary and neutral representation could also be used to achieve enhancement to many CAPM software packages of similar structural design.

## 4.3 DESIGN CRITERIA FOR SYSTEM-WIDE DATA REPOSITORY

As part of the information architecture proposed, a new approach to consolidating data is offered. The approach was conceived after having recognised the need for solutions in the following problem areas (as highlighted in Section 2.2.2):

(a) lack of adherence to any standard architectural models of information, which undermines the semantic integrity of shared information, can result in invalid combinations of data (this as a consequence of incompatibility and inconsistency in data definition and format);

(b) multi-database concurrency problems, related to database transactions and concurrency control and a means of consistently handling data access, transfer and presentation issues, for data which is normally fragmented and distributed across various database. This includes

- independent and transparent data access, i.e. access without having to specify or know where the data is stored;

- support for multi-user access where appropriate in-built system mechanisms ensure that competing applications wishing to access data entries do not endanger the integrity of the data repositories.

**Figure 4-4** System-wide data repository

The data repository or 'data warehouse' comprises a directory of shared elements where common data definitions are recognised throughout the enterprise as global data elements, i.e. they have globally understood inter-relationships. The system-wide data repository would serve as the focal point for access of shared data and provides users and software applications with a consolidated view of information — one that is independent of physical media or data location; refer to Figure 4-4 for an illustration of the concepts involved here.

It should be noted that data which will always be unique to an application can remain in its own private local database to realise efficient processing. By restricting the number of information entities made globally available there will be a diminution in the amount of data exchanged, thereby reducing network traffic and bottlenecks relating to data access.

The data unification mechanisms of the information architecture conceived unifies the use of a distributed set of data repositories which will be located over several computer nodes connected over a local area network.

The following techniques were proposed and advanced to help overcome the problem specified in (a):

- apply the generic reference models identified and defined by the author to offer a degree of formalism in terms of recognising information of common concern over an extended domain;

- adopt a suitably defined logical data model (to be further discussed in Section 4.3.1) to represent information entities and attributes in a uniformed manner where these entities comprise information fragments within distributed data repositories. This can

  - alleviate problems of incompatibility and inconsistency in data definition and format;

  - facilitate required changes to the data model and physical data.

The IIS is utilised to address some of the inherent problems referred to in (b). It would provide software applications with structured access to common integration services for communication and management of data stored in distributed data respositories. Database 'drivers' would enable the following capabilities to be offered via the IIS:

• connection and direct communication with a number of proprietary databases (in a manner which is independent of the physical location of the information fragments involved); and

• a means of accessing information held in the database.

It should be noted that besides providing services for database queries, the 'drivers' must also be able to support data-intensive activities which require large amounts of data to be accessed, transferred and processed.

## 4.3.1 A Logical Database Model of the Data Repository

A relational data model is recommended as an underlying structure for accessing information fragments from the distributed data repository. In a relational model entities, relationships and attributes are represented in the form of two-dimensional tables known as relations. Records are assimilated into the rows of the table and each set of attributes forms a column. In a relational database entities are stored totally independently, i.e. the existence of a relation is not dependent on any other relation. Logical associations among the stored data are exploited through relational operations such as select, project and join which can be used to create new tables. Any number of operators and relations can be combined in a 'relational expression' and used to answer almost any query. The entities, attributes and relationships of the conceptual data model can often be modelled directly as relations in a relational database model.

The use of the relational model rather than hierarchical or network models demands less of a compromise when representing and transforming real-world data relationships. This claim is because of the following:

(a) In many real life situations, relationships cannot naturally be represented by a hierarchical model (where normally one-to-many segment types are used to represent successive levels in the hierarchy, thereby relating entities to one another). It is not easy, for example, to directly represent relationships among segment types at the same hierarchical level nor is it possible, without introducing data duplication, to represent many-to-many relationships between entities.

(b) Network structures offer greater scope in representing data relationships when compared with hierarchical structures, albeit at the expense of simplicity. This is certainly true of their underlying physical storage structure. The need to transform many-to-many relationships by the construction of a network model results in the need to make more or less irreversible decisions about the nature of relationships between entities when the data model is designed. It should be noted that the network model, whilst permitting means of representing many-to-many relationships without introducing duplication of record occurrences, does make retrieval of data a laborious process.

The importance of the relational model is widely acknowledged and the development of Relational Database Management Systems (RDBMS) is progressing rapidly. Potentially, the relational data model offers the following advantages:

- **Ease of Use** Visualisation and clarity of data, which are represented in two-dimensional tables, is better for both programming and non-programming users.

- **Simplicity** Here all data is viewed in the form of relations (tables), thereby allowing easy data manipulation and query via Structured Query Language (SQL).

- **Flexibility and Relatability** With relational operations a standardised and effective way of decomposing and recomposing relations is provided. This approach enables the incremental building of larger systems module by module.

- **Security** Security controls can be easily implemented where security authorisation will relate to relations to protect company sensitive attributes.

- **Data Independence** There will be a need for most databases to grow by adding new attributes and new relations and also for data to be used in new ways. The relational model supports dynamic reorganisation (i.e. extension and modification to the structure) of the database without affecting existing applications. This is important because of the excessive and growing costs of maintaining the software applications of an enterprise and its data from the disruptive effects of database growth.

- **Data Manipulation Language (DML)** One of the strengths of the relational model is that it is generally supported by high level non-procedural, set-oriented languages, such as 4GLs (fourth generation languages), to enable flexible access, management and presentation of data stored in the database.

As illustrated in Figure 4-4, the 'three-schema' architecture was adopted in the author's approach in order to map from the logically integrated relational model to the physically distributed database in which the actual data is stored. The three schema approach requires definition of the following information schema:

- **Internal Schema** This represents the physical organisation and storage of the information.

- **Global (or Conceptual) Schema** This represents a composite view of a common pool of shared data. The objective is to provide a consistent definition for meanings and inter-relationships between data entities in order to aid information management.

- **External Schema (or Local Viewpoint)** This describes the use of information, i.e. the information required by a user or an application. Objects in the external schema are automatically mapped onto information attributes in the internal schema via reference to the global schema.

The shared elements represented by the global schema correspond to information entities and attributes which are defined and specified in the generic reference models. Hence with the three schema approach all commonly shared physical data which is fragmented across the various local database of the data repository can be flexibly mapped and associated through established object relations. Thus each software application only requires one interface to the global schema, thereby enabling access to a common pool of information. Similarly, each local data source only requires one interface. Changes are required to the global schema if new data entities are provided or old entities are removed. In the event of a physical restructuring of the database, reorganisation only affects the internal schema and a single mapping between the global schema and the internal schema. Mappings between the external schema and the global schema remain unaffected.

Hence the three schema approach provides a greater degree of data independence when compared with the two or single schema approach where information is typically embedded. This causes them to be exclusive in nature with regard to the functions concerned, thereby making it rather difficult to separate and access them. The use of a three schema database architecture leads to improved flexibility where changes can be made without the need to modify applications. Individual local database can retain their autonomy, with a focus on serving their existing customer set.

## 4.3.2 Future of Relational Database Management Systems

It would be useful at this stage to provide some insight into future development of RDBMS. This provides a justification for choosing the relational model as a viable long-term basis for manufacturing systems integration which can be expected to cope with advances in application requirements and database technology.

Traditionally, RDBMS have lacked adequate data types to fully represent engineering data. In a typical manufacturing company, engineering data will include CAD/CAM generated product models, part drawings and tool paths (used for generation of CNC programs). This engineering data is highly complex, where for example, relationships between the lines and angles of their vector graphics must be effectively represented and preserved. Within the literature available, it is strongly advocated that such data be best handled and supported by some form of object orientation.

Progress is being made within the database research community towards extended RDBMS so as to provide increased functionality and support for object concepts. Examples include Intelligent SQL, Objects in SQL, and current efforts of the ANSI X3H2 committee (SQL3). Indeed major RDBMS manufacturers such as ORACLE, Sybase, Informix, Borland and Ingres have begun to add extensions to their products, which can handle simple forms of object orientation to support complex and user defined data types which include:

- provision of large data fields that can store binary data;
- inclusion of stored procedures and triggers which allow for the storage of data along with programs and procedures that apply to them.

Choice of relational technology should provide a foundation for the future adoption of developed and enhanced forms which can also benefit from the technology's inherent simplicity (which has been recognised as one of the great strengths of the relational model). In the long term there exists the prospect of managing both engineering and non-engineering data as object-oriented relational database technology matures and becomes readily available. Such requirements have been identified and considered in specifying the conceptual solutions proposed.

### 4.3.3 Database Access Approach

Depending on the class of user interacting with a database management system, three access approaches can be adopted for relational database (Figure 4-5):

**Database Query Language Facility**

RDBMS essentially adopt a table based organisation of data. The user must establish links from table to table, thereby making tables behave temporarily as relations. This can be done through the provision of fourth generation database query languages such as SQL. In various forms SQL is offered by different database manufacturers, for example, Query DL/1 for DL/1 (IBM), SQL for DB2 (IBM), SQL for ORACLE (Oracle), NATURAL for ADABAS (Siemens), SQL for INGRES (Ingres), and so on. Via the help of a basic structure, based on the following three key words, SQL allows complex queries to be made:

> **SELECT** Fields (columns) to be displayed;
>
> **FROM** Relations (files) containing the fields;
>
> **WHERE** Conditions (giving the selection criteria).

The use of SQL is appropriate for non-programming users as it enables independent access to information necessary for the particular area of application concerned.

**Database Programming Language Interface**

This approach is well suited for the programming user. Special database interface instructions (i.e. embedded SQL), which are normally offered as a set of readily available program functions, are incorporated into the application program to enable access to the database through the database management system. However, when using this approach, if changes to data access, manipulation and presentation are required, a change to the application program(s) is necessary.

### Data Access Tool

Despite the widespread commercial and industrial use of SQL, two major technical barriers currently exist with respect to data access across multi-vendor RDBM. These two barriers are:

- **SQL Language Differences** In reality each vendor has their own SQL 'dialect' where each is peculiar to the RDBMS it supports. It has been estimated that approximately 80% of the SQL syntax and semantics is identical between dialects. However, the remainder includes subtle differences which make it extremely difficult to convert from one SQL dialect to another.

- **Differences in Message Formats and Communication Protocol** In practice different RDBMS vendors have targeted the use of their products at different hardware platforms and operating systems. As a result there exist distinct differences in the message formats and communication protocols of the different RDBMS which can further accentuate the heterogeneity of integrated systems. This undoubtedly causes interconnection problems between multi-vendor RDBMS.

The emergence of data access tools, such as DataDirect Explorer and UNIFACE offered by INTRESOLV and COMPUWARE respectively, is much awaited. They can to a certain extent help resolve problems associated with SQL language differences and differences in message formats to accomplish access to a distributed and heterogeneous set of data repositories. The data access tool provides capability for data access, analysis and presentation. It is designed to provide transparent access to a number of databases using a single consistent interface. For example, DataDirect Explorer contains database 'drivers' which are Open Database Connectivity (ODBC) compliant to support more than thirty-five DBMS which range from flat-files to relational database systems such as Sybase, ORACLE, INGRES, Informix and DB2. ODBC[1] is a specification published by Microsoft and is rapidly becoming the industry standard for database independent data access. The data access tools make it easier for

---

1. **ODBC** allows developers to design applications based on a common API rather than a specific DBMS, thus creating applications that are easily portable from DBMS to another.

a broad range of users to access, query and analyse, integrate and report mission-critical information for improved decision-making. The use of such tools represents a new approach to the process of accessing and understanding relational data where the end users are not required to know cryptic table names, column names or even complex SQL statements that accompany direct access of a database.

## 4.3.4 Database 'Driver' Requirement for IIS

Here the database 'driver' is aimed at enabling consistent, reliable, transparent and open access of information stored in the database system by the distributed software applications via the IIS. When implementing the database 'driver' to enable structured access (via common integration services for communication and management of data) to fragments of information stored in distributed data respositories, it is essential for the 'driver' to:

(a) bring the database system up to a level of conformance which allows it to utilise the integration services of the IIS; and

(b) enable a degree of tailoring to account for the idiosyncrasies of the relational database management system concerned.

As illustrated in Figure 4-6, the 'driver' is required to operate at the interface between the IIS and the database system. It provides a set of services to link the IIS and database system; it handles the IIS service requests and the inherent complexities of the database system concerned to access the information required.

Generally, the database 'driver' comprises the following:

• **IIS Interface Module** to provide an interface to the IIS to bring the database system serviced to a level of conformance which facilitates utilisation of the common integration services (i.e. information management and access) offered by the IIS. This is to enable transparent access to information (which is held in the distributed data repositories) from the business applications via the IIS embedded communication protocol.

• **Database Interface Module** to provide an interface to a specific database system, configured to translate between IIS and proprietary (i.e. specific to the database system concerned), message and data formats as well as SQL dialect requirements. Typically, the database programming language interface of a given RDBMS (i.e. embedded SQL) is used to access the database system.

**Figure 4-6** Database 'driver' interfaces to data repository and IIS

The database 'driver' should perform the following set of tasks when it is initiated by the IIS:

- establish connection to the database system concerned;
- route information access requests received from the functional modules (via the IIS) to the database system. These requests are converted and conform to the SQL compliant services provided by the 'driver';
- receive the requested information from the database, format it and either send it back to the functional module or present it to the end user to view;
- report and recover from errors which may occur.

Thus the database 'driver' acts as a standard interface to connect RDBMS from different vendors where its underlying services can be utilised by any functional component requiring information access to the database via the IIS. It should be effective in supporting data intensive activities which normally require large amounts of data to be accessed, transferred and processed.

## 4.4 SUMMARY

In relation to the overall methodology conceived and advanced to enable integration and interoperation of functional modules, this chapter has a direct bearing on the following three sub-methods:

- **Specification of Reference Models** A set of widely applicable generic reference models representing prime information of common concern among functional components has been identified and described in order to help structure and facilitate integrated manufacturing system design.

- **Use of an IIS** The IIS, which assumes responsibility for maintaining knowledge of integration details, is used to simplify problems of realising interconnection among functional modules. It resolves differences in the physical system relating to heterogeneity, distribution and data fragmentation.

- **Interfaces to Physical Resources** Database 'drivers' are needed to effectively support data intensive activities. The 'driver' enables the following:

  - brings the database system serviced to a level of conformance which facilitates utilisation of the common information management and access services offered by the IIS; and

  - flexibly maps software applications onto system resources, i.e. achieves a mapping between physically distributed database and the logically integrated relational data repository which corresponds to the ANSI SPARC three schema architecture.

# 5 Integrating Infrastructure to Underpin Interoperation

## 5.1 FUNCTIONAL INTERACTION REQUIREMENTS

The identification and specification of generic reference models, to create synergy between functional components through shared information, is merely an entry point towards achieving software interoperability. In order that the benefits of software interoperability can be more fully realised, the next crucial step is to address problems of functional interaction among functional modules. In practice, this demands co-operation among interoperating software components to establish well defined communication channels which collectively promote and enhance intra-organisation interaction and co-ordination of activities. Hence an inherent capability to control system behaviour is essential. Indeed this is one of the requirements which needs to be satisfied to enable software interoperability in an effective manner.

In order to provide a generalised and flexible way of enabling functional interaction between the components of integrated manufacturing systems a framework is required which formally structures and manages functional interaction among functional components. In this way a group of normally autonomous functional components can function as a co-ordinated whole, with means of maintaining discipline among them. The following requirements are identified as being necessary to enable and support functional interaction:

1. **Establish Association Between Functions and Their Required Information**
   Typically in a manufacturing company, the sequence of activities necessary to perform part manufacture in progressive stages, relies heavily on their functional dependencies and on information needs between the different stages. Generally, the association between functions and their required information is not formally and clearly defined, particularly across different functional domains, and much depends on the users to ensure control and co-ordination of systems as a whole. Consequently, this contributes to significant delays in transactions as well as misunderstandings and conflicts. The existence of a formal association, established between functions and their required information, is viewed to be highly beneficial in terms of providing means to structure and govern system behaviour (during run-

time), where associated preceding and succeeding activities and consideration for their shared information needs (as defined by the generic reference models in Section 3.3.1) have to be taken into full account.

In addition, any associations established between functions and information must be configurable in nature. This is necessary to suit both specific and changing functional and information needs of users.

## 2. Capability for Controlling and Co-ordinating the Functional Activities

Consideration needs to be given to the availability of shared information necessary to satisfy and support the (run-time) activities carried out by functional modules. This needs to be done with close reference to associations established between the functions and their required information, as stipulated in (1). Mechanisms will be required to closely monitor the availability of such information (held in the data repository) and to initiate or trigger appropriate functional activities in need of that information for further processing.

## 3. Means to Effectively Coalesce all Interacting Functional Components

It is necessary to provide easy user access to associated functions which may be distributed across the local area network so that human-centred tasks can be appropriately supported. Such an approach can provide users with a global perspective when conducting their specific tasks so as to ensure better and more informed decision-making.

## 4. Means of Making Reference to Engineering Data

Normally in a manufacturing company, some design- and product-related engineering data, which typically will include part drawings and product models, is frequently referenced to help determine resource requirements (e.g. toolings, fixtures, etc.), plan routes for part manufacture, select part inspection routines and so on. Controlled access to this engineering data can prove very useful and can effectively contribute globally towards more informed decision-making and planning. Thus it would certainly be helpful to have knowledge of the existence of such data and where it can be accessed or requested within the system. Recognising the current limitation of RDBMSs in not being able to physically handle engineering data (as discussed in Section 4.3.2), the task, therefore, is to provide a means whereby the user can at least be informed of its availability and physical location over the local area network.

It is expected in the near future that this task requirement could be further enhanced by incorporating capability to manage engineering data, as suitable enabling technology needed to co-manage both engineering and manufacturing data becomes more readily available. It would then be possible to establish relationships among information entities associated with the respective functional modules with relative ease.

Increasingly, Product Data Management (PDM) systems for workflow management of product-related data are becoming available, such as IMAN, Optegra, Metaphase, Sherps, WorkManager from EDS, Computer Vision, Structural Dynamics Research Corporation, SHERAP and Hewlett Packard respectively. They link all engineering related data into a single enterprise-wide cohesive system so as to ensure complete and consistent product information presentation to the end user. Work is on-going among major vendors to establish links between their PDM systems and other functional modules, particularly production planning and management systems.

## 5.2 CONTEMPORARY SOLUTION: FUNCTIONAL INTERACTION

A product literature survey was conducted and it was found that there are very few solutions available which have been designed with the purpose of achieving functional interaction, particularly between legacy applications from different sources. Products such as Systems Integration Manager (SIM) from Manufacturing Systems Portfolio PLC (UK) have been produced and used by systems integrators to tie together a set of software applications into a coherent system. To help assess the capabilities of contemporary commercial solutions to the functional interaction problems, the properties of SIM, which was chosen as a representative proprietary product designed to realise functional interaction, were investigated. The properties considered were its scope of functionality and level of effectiveness in facilitating functional interaction (with relation to the requirements identified in Section 5.1). This study of an example product was beneficial in terms of providing a general indication of the focus and usefulness of a class of contemporary solutions aimed at enabling and supporting functional interaction.

## 5.2.1 Overview of Systems Integration Manager

Messages are the core of the SIM system, realising interoperation between applications via message passing. SIM provides the following features:

| Features | Description |
|---|---|
| MESSAGE PASSING | The routing of messages between applications. |
| MESSAGE HANDLING | The definition and validation of messages. |
| EVENTS AND ACTIONS | The specification of the occurrences of conditions such as the receipt of a given message and the action to be taken. |
| PROCESS CONTROL | The administration of applications. |
| FILE TRANSFER | The transfer of text files between computers. |
| STATISTICS | SIM maintains statistics on the frequency of service requests. |
| PRINTING | The printing of files. |

**Table 5-1** Features of SIM

All these features are available as a set of ANSI-C program functions and need to be incorporated into newly developed and existing applications in order to enable an interface to SIM to be established, thereby enabling use of the facilities it provides to allow interoperation.

The behaviour of SIM is described, defined and governed through a set of configuration tables (Figure 5-1). In order to be incorporated into the SIM approach, the configuration tables need to comprise the following:

- Registration of applications which includes the assignment of a unique identifier for each application when specifying its actual physical location for direct access.

Example

| | |
|---|---|
| Event ID | 21 |
| Message Receipt | 5 |
| Action ID | 10 |
| File Transfer Request ID | 260 |

**Figure 5-1** Interoperation between applications enabled through SIM

- Predefinition of the structure of messages (which includes its type, length and assigned identifier) and a specification of its targeted destination.

  Example

  | | |
  |---|---|
  | Action ID | 10 |
  | Application Loaded | 12  ◀— Application ID |
  | Application Unloaded | 16 |
  | Application Terminated | 14 |

- Specification of 'events' which can be described as some expected happening. For example, receipt of a message or the transfer of a file.

  Example

  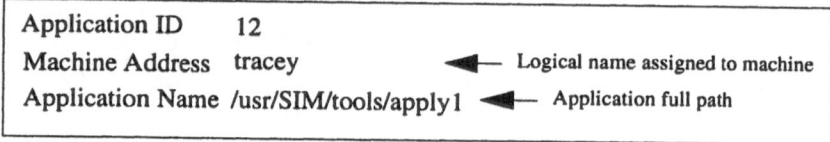

  | | |
  |---|---|
  | Application ID | 12 |
  | Machine Address | tracey  ◀— Logical name assigned to machine |
  | Application Name | /usr/SIM/tools/apply1  ◀— Application full path |

- Definition of 'actions' that should be performed when an 'event' happens.

  Example

  | | |
  |---|---|
  | Message Type | ASCII |
  | Message Text | Generate Schedule report |
  | Message length | 52 |
  | Destination Application ID | 12 |
  | Message ID | 5 |

As illustrated in Figure 5-1, during normal operation incoming messages from applications are received and queued by SIM. They will be validated and processed with reference to the configuration tables. In response to the input received, predefined 'actions' and corresponding output messages will be activated and delivered respectively by SIM to trigger off targeted applications.

## 5.2.2 Focus and Limitations

In relation to the requirements identified for functional interaction, the following highlights the characteristics of the contemporary solution studied:

- **Event Driven to Control and Co-ordinate a Sequence of Activities** The solution is basically event driven and is loosely based on predefined message passing between applications. It does not take into account information resources required to drive and support the various activities within the system. Thus association between functions and information is not considered and availability of required information is not monitored when attempting to validate and control the sequence of activities. Rather, system behaviour control (during run-time) is achieved merely through management of messages.

- **No Effective Means of Coalescing Interacting Applications** All applications concerned need to directly incorporate special routines within themselves so that they can utilise the interaction facilities and services provided. Hence the applications concerned are rigidly configured as the interaction knowledge they require will be embodied within each one of them. Such solutions will face similar problems to 'pair-wise' integration (as mentioned in Section 2.2.1) where the complexity of such systems will grow substantially (theoretically in a square law fashion) as the number of interconnected applications grows. As a result, it will be tedious and cumbersome to manage changes to any of the interacting applications within the system, and it may not be practical to give due consideration to the full implication of changes to the system as a whole.

Although the application of SIM might be satisfactory in facilitating interaction and co-ordination between isolated 'islands of computerisation' in an event-driven manner, the following additional requirements are deemed to be necessary when dealing with enterprise-wide integrated manufacturing systems:

- Control and co-ordination of the work flow (i.e. system behaviour during run- time) should not only be event driven but data driven as well. This is to ensure that the use of information resources, particularly those sharing information of common interest, are considered during functional interaction. This is essential in order to properly and accurately validate and co-ordinated the sequence of activities in relation to the availability of information set in the context of realising more global goals.

- Simplify interconnection between interacting applications via an IIS. This is to provide common integration services to all conforming applications, via use of common integration services provided by an infrastructure.

## 5.3 FUNCTIONAL INTERACTION MANAGEMENT SERVICES

In order to provide the capability needed to meet the identified requirements to facilitate functional interaction management a set of services which serves as an application enabler is required to help formally structure and facilitate functional interaction in a controlled and deterministic manner. Collectively, they can be viewed as constituting high level services and need to built upon the low level, general purpose integration mechanisms and tools of the IIS (as defined in Section 3.4) which provide standard data inter-communication and information transfer facilities to interconnect software components.

In this section the scope and functionality essential for the following set of services, as illustrated in Figure 5-2, will be discussed in general. In addition, they can also be compared to existing contemporary solutions of such nature to identify their limitations and inherent constraints.

**Figure 5-2** Framework for functional interoperation

## Function-information Association Management

As illustrated in Figure 5-3, predefined relationships between functional activities and their shared information (which associates information entities to the relevant functions, designating them as either an input or output requirement or both) need to be formally established and clearly identified. The information model, which encodes relationships between information entities, corresponds to the generic reference model (see Section 3.3.1). The function-information association which can be represented in a relational table will be referenced (during run-time) in a way which governs system behaviour in a controlled and co-ordinated manner, i.e. in accordance with predefined relationships established between the information entities and the functional activity concerned. Use of this table-driven approach offers the following benefits:

- a natural means of assigning information input/output requirements following front-end system design based on modelling and analysis; and

- information association and hence inherent traceability and accountability to the functional activity.

**Figure 5-3** Function-Information Association Table to support functional interaction

This table can also serve as an intermediary storage and representation facility which can be populated with data generated from the system's design model. Later chapters of this book illustrate the adoption and use of a methodology and software tools to enact information and function models in this way. This approach was conceived in order to establish and configure necessary run-time associations in a highly flexible manner, through identification and representation by the Function-Information Association Table.

## Interaction Management

This is responsible for controlling and co-ordinating work flows associated with functional activities via status management and transaction control. Transactions which involve (a) initiation of interoperating functional activities, and (b) request and exchange of shared data between interoperating functional components are closely monitored during run-time in order to validate:

- that the required sequence of functional activities has been performed;
- appropriate associations between functions and their required information are maintained; and
- that the integrity of shared data in the data repository is maintained.

Status markers and triggers need to be implemented to provide a mechanism to control and co-ordinate functional interaction. Here, status markers can either be assigned automatically or interactively by an operator (in response to certain system requests to alter them during normal operation), where they can subsequently be used by the system to trigger or block further transactions between interoperating functional components.

To enable concurrency, i.e. parallel operation of functional components, the operational status of the threads of functionality embedded within each functional module is to be monitored closely; for example, order registration and resource management activities which constitute production planning are clearly defined and handled separately. Status markers are also used to reflect changes of the activity instance, thereby effectively controlling and co-ordinating the start-up and shutdown of subsequent and dependent activities.

**Figure 5-4** Overview of interaction management of functional activities

Interaction management would need to facilitate the following services:

- update the (a) processing status of functional activities (i.e. to reflect whether work is in progress, completed or pending for the particular part or order being worked on), and (b) instance to reflect the processing phase at which the part is currently at;

- search and view facility to indicate status and instance for the part(s) being processed; and

- job loading for the functional activity concerned.

Refer to Figure 5-4 for an illustration of the operational view.


**Engineering Resource Management**

This offers a means of managing engineering data which includes product models, part drawings and computer numerical control (CNC) programs necessary to support part manufacture. However, it does not physically process the engineering data but effectively serves as an archive where these resources are registered and cross-referenced based on an agreed convention, such as part number as illustrated in Figure 5-5. Via resource management the end user can be informed of availability and physical location of engineering data, distributed across a number of computer systems.

**Figure 5-5** Engineering resource management

The following services should be made available for engineering resource management:

- registering the source location of the relevant engineering data; and
- search and locate facility, for example, with reference to the function responsible for managing it.

Today there are a number of commercially available software packages dedicated to engineering (or product) data management. However, they tend to be exclusive for the CAD/CAM domain and very much proprietary in nature in terms of the format and structure in which the engineering data are to be stored and managed. As the engineering data management software package is responsible for handling engineering data and if found suitable it can effectively serve as the back-end support to administer and manipulate the engineering data when coupled with engineering resource management facility. The amount of work involved to establish the required interface would undoubtedly depend on the degree of openess of the software package concerned.

## Configurator

This was conceived to serve as a tool to enable configuration to suit specific user needs. It facilitates the addition, deletion, and display of functional activities, which are to be managed, as well as enabling the attribution of data to entities in the information model (which corresponds to the reference model as described in Section 3.3.1). It should provide editing services which would allow relations between the functional activities, information entities, engineering resources and activity instances to be altered quite easily. Its functional structure and scope are illustrated in Figure 5-6.

The configuration information describes the working parameters necessary to drive and control functional interaction; during system run-time this information is referenced by interaction management and engineering resource management. In order to ensure consistency and integrity of the required configuration all alterations or reconfiguration are to be done off-line and when the functional activities managed are inactive.

**Figure 5-6** Structure of configurator

The following data which is relevant to the normal operation of functional interaction is to be stored locally in a database:

- attributes of the functional activities and information entities to be managed;
- type and nature of relationship between functional activities and information entities;
- operational status of functional activities supporting part manufacture; and
- a description of the logical links to the engineering resources based on their designated call names.

In addition, it would be beneficial to include error recovery facilities for the configurator to help report, troubleshoot and recover from possible error situations.

## 5.3.1 Distributed Functional Interaction Management

The required functional interaction service can be directly invoked if the database (populated with the relevant working parameters to support functional interaction) resides locally on the same host computer as the functional activity. However, in order to support functional interaction in a distributed computing environment, where the functional activities are spread over several host computers connected via a local area network, remote versions of the services are necessary. This is to allow any host computer connected to the network to have access to the functional interaction services. As illustrated in Figure 5-7, in order to enable distributed functional interaction management communication mechanisms must be implemented to connect between the remote host computer where the functional component resides and the computer system where the set of functional interaction services resides.

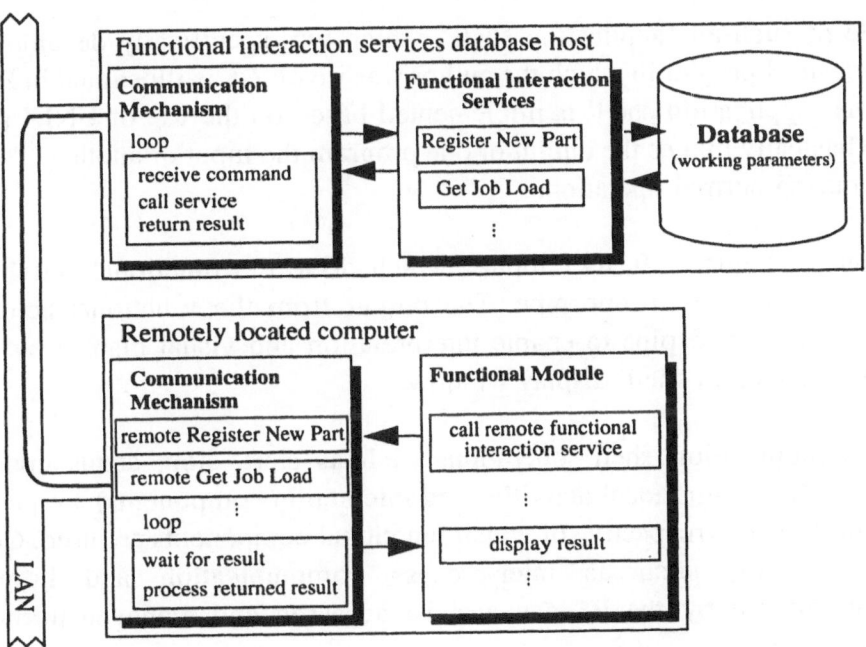

**Figure 5-7** Communication mechanism to support distributed functional interaction

The following capabilities are necessary to enable distributed functional interaction management:

- requests made by the functional activities to be relayed to the host computer where the set of functional interaction services reside for processing to occur;

- request made by functional activities for the service(s) to be interpreted;

- functional interaction service(s) to be invoked as required;

- ability to relay responses or outputs from the functional interaction services back to functional activities requesting its service.

## 5.4 REQUIREMENT FOR FRONT-END USER INTERFACE

As part of the working framework to enable interaction among functional components over the IIS, the need for a highly reconfigurable and hence widely applicable front-end user interface or 'application shell' was required. It provides easy access to associated functional activities which are normally distributed within the system.

An example of such an 'application shell', which was specifically developed for a previous industrial project in which the author was involved, is illustrated in Figure 5-8. Here the 'application shell' is implemented based on the use of UNIX pipes. A UNIX pipe basically makes the output of one program the input of another. Two pipes are created during normal operation.

As illustrated in Figure 5-9, user inputs, which drive the functional activities, are emulated and redirected to one pipe. The output from the functional activities is redirected to the second pipe to enable interpretation and visual display by the end user via the 'application shell' display window.

Through the 'application shell' operations such as start, stop, status update and queries (for both the functional activities and interaction components) are performed in a standardised way irrespective of which functional components are used. Common integration services, such as inter-process communication and information management, offered by the IIS can also be accessed and managed through this standard interface.

When designing the front-end 'application shell' it is essential to satisfy the following requirements:

- To provide end users with an abstracted working viewpoint which supports them in an effective manner when conducting their specific tasks. This satisfies the need for an end user interface capability which reflects inherent human interaction requirements (as stated in Section 1.4) so as to enable software interoperability in an

Figure 5-8 'Application shell' to coalesce interoperating functional components and integrate with functional interaction and IIS services

**Figure 5-9** Communication interface via 'application shell'

effective manner. This 'standard' human interaction facility would provide end users with a global perspective which enables better and more informed decision-making in relation to performing their specific tasks.

- To make functional interaction management easier by effectively insulating the user from complexities and details involved in the underlying interoperation processes which will be taken care of by the relevant functional interaction and IIS services offered.

## 5.5 SUMMARY

The following set of functional interaction services has been conceived with the purpose of (a) providing a framework which formally structures interaction among functional components, and (b) enabling the flexible integration of functional components:

- interaction management
- engineering resource management
- function-information association management
- configurator

Restated the requirements for functional interaction illustrates a need to:

- **Establish Association Between Functional Activities and Their Required Information** These formal associations can be represented in the form of a function-information association table which identifies predefined relationships between information entities (which correspond to the generic reference model described in Section 3.3.1) and activities performed by the functional modules. This table is to be referenced (during run-time) in a manner which governs system behaviour in a controlled and co-ordinated way. Here associated preceding and succeeding activities and consideration for their need for shared information is taken into full account.

  The configurator was conceived to facilitate flexibility in a manageable fashion, where via its editing services, the functional interaction services can be configured quite easily to suit specific user functional and information needs.

- **Establish a Capability for Controlling and Co-ordinating the Sequence of Functional Activities Performed in a Distributed Manner** The interaction management service is responsible for providing this capability. It incorporates mechanisms to:
  - monitor the availability of shared information via its status management capability;
  - initiate or trigger appropriate functional activities which are in need of that information for further processing.

- **Establish Means of Effectively Coalescing the Interaction of Functional Components Via a 'Standard' Human Interface** A generic 'application shell' to enable human interaction with functional components via the functional interaction and IIS services is essential. The 'application shell' effectively serves as a front-end user interface for functional interoperation, where easy access to associated functional activities which are distributed throughout the system at different network nodes is offered. This provides the end user with a system-wide working viewpoint of the effect of conducting their specific tasks. The 'application shell' is responsible for drawing together various interacting functional components in a manner which leads to synergy between them. As a result it promotes and enables co-operative as well as better and more informed decision-making across the enterprise.

- **Establish Means of Referencing Engineering Related Data** It can be achieved via the Engineering Resource Management service conceived. This effectively serves as an archive for engineering related data where end users are informed of the availability and physical location of information entities which will be distributed as information fragments stored in different database at a number of computer nodes connected to a local area network.

The set of functional interaction services overcomes some of the limitations inherent in contemporary solutions by offering the following advantages:

- **Control and Co-ordination of the Functional Work Flow in an Event- and Data-driven Manner** Here due consideration is given to the availability of shared information and the need by the functional activities concerned in order to properly and accurately validate and co-ordinate the sequence of activities performed on a system-wide basis. With reference to the association established between functional activities and their required information, concurrent operation of functional activities is made possible in a flexible but controlled manner (via the interaction management service). This enables interacting software applications to run in parallel and their activities to be synchronised based on the availability of their required information which is monitored by appropriate mechanisms provided.

- **Simplify and Manage Interconnection Between Interacting Applications Via Use of a Domain-specific IIS** The set of functional interaction services is built upon the IIS in order to utilise the more general integration services provided which includes inter-process communication and information management. This is to help simplify interconnection between applications by delivering these general services in a form which is more suited to typical users found in manufacturing organisations. Furthermore, interaction knowledge is embodied in the functional interaction services which is used at run-time to integrate functional activities with human interaction enabled over the IIS, via the generic 'application shell'. Thus it is only necessary to link the applications concerned to this front-end common end user interface in order to make use of the services provided. This essentially removes the need to incorporate interaction knowledge in each individual application, thereby simplifying interconnection, facilitating interaction and providing opportunities to standardise and modularise functional components.

# 6 System Life-cycle Support

*The only constant is change.*
Bishop

## 6.1 REQUIREMENT FOR INTEGRATED LIFE-CYCLE SUPPORT

In reality integrated manufacturing systems are characteristically evolutionary in nature and must be adaptable and responsive to changing needs. For example, a very common requirement is further integration with other functions and a re-engineering of existing functions in order to modify and enhance the process capabilities of any given enterprise. Change is essential in order to provide competitive differentiation and to ensure the survival of companies in the face of changing customer, supplier, financial and labour markets. As Tom Peters (1989) aptly pointed out in his highly publicised book entitled *Thriving on Chaos,*

> *To survive and become superlative in today's economic environment, the flexibility to react and be responsive to changes is highly desired ... Impermanence is a cherished quality for excellent firms.*

Change in the requirements and characteristic properties of integrated manufacturing system will inevitably affect dependency relationships and information flows among interoperating functional modules in the system. Hence the author strongly advocates the need to effectively support an integrated system through its life-cycle, which will involve the following phases:

• **Conceptual Design** The prime focus is deciding what a system should do.

• **Detailed Design and Implementation** This involves specifying how the global requirements defined can be realised in terms of building the required solutions.

• **Operation and Maintenance** This characterises the working life of the installed solution, as well as necessary adjustments and repair during the operational lifetime of the system.

In order to facilitate ease of system development and change management, it is necessary to share and channel usable results and data between the different life-cycle phases in a consistent and accurate manner. However, there is presently an absence of an integrated, formalised and structured approach which straddles the various life-cycle phases, that can help support systems as they evolve. As is evident from the literature survey, no one methodology includes a capability for modelling the functional, information, dynamic and decision aspects of integrated manufacturing control systems. As a result, independent and separate use of a number of methods will be required if the formal modelling and construction of systems is required.

As previously discussed the formal modelling of systems can provide an entry point for supporting the life-cycle of manufacturing systems where the models created (of function and information aspects) can serve as a source of knowledge during different life-cycle phases. However, additional life-cycle support tools coupled closely to the modelling tool are needed. Such a software toolset should exploit the knowledge contained within the model in order to:

- reference the functional activity-based and information models created during conceptual and detailed design;

- ensure compatibility and continuity between different life-cycle phases, i.e. maintain consistency between models produced and used at each life-cycle phase; and

- control and enforce structured implementation and change management.

Therefore, an integrative approach is strongly advocated by the author to facilitate life-cycle support of systems which involves the design, implementation, run-time and change processes. In this approach, system design and modelling methods which typically provide a means of representing activity-based, information and behavioural views of a system serve as the entry point. The information and function models created are exploited in downstream life-cycle phases to ensure clarity, consistency, accuracy and re-utilisation of knowledge and data between phases.

## 6.2 CASE STUDY: REALISING THE INTEGRATIVE APPROACH

In this section the integrative approach conceptualised by the author to facilitate life-cycle support is discussed. As part of the approach the IDEF methodologies, namely $IDEF_0$ and $IDEF_{1X}$ to enable functional activity-based and entity-attribute relationship modelling respectively were chosen as the starting point (refer to Appendix III for an overview of the IDEF methodology). They were chosen for the following reasons:

- They had demonstrated their usefulness as 'simple and effective communication tools' which encourages end user involvement as well as co-operation with system builders.

- They have a growing popularity and acceptance, this being evident from the significant levels of research and industrial applications published in the literature. This reflects primarily their accessibility and potential in a wide range of applications.

- They are able to provide comprehensive functional activity-based analysis capability.

- These modelling methodologies offer opportunities for enhancement and integration with other tools.

- Although the normal application of $IDEF_0$ and $IDEF_{1X}$ are separate and independent of each other, they are able to offer a formal and relatively complete representation of the manufacturing systems from their different modelling perspectives (i.e. function and information).

- $IDEF_{1X}$ is a natural choice for representing a relational information model, as the model created has an inherent one-to-one correspondence with the entity-attribute relationships normally defined in a relational database.

Figure 6-1 provides an overview of the system modelling and implementation environment conceived to enact static functional activity-based and information models. The purpose of this environment is to provide a formal and structured approach to:

(a) facilitate implementation processes based on specific end user requirements concerned with information systems comprising relational data repositories;

(b) enact the information and functional activity-based models created in order to establish and configure associations, as identified and represented by the function-information association table to support the functional interaction services. This table represents and encodes system behaviour.

**Figure 6-1** Software toolset for integrated life-cycle support

The environment itself is built from an aggregate of the following tools:

    (a) $IDEF_0$

    (b) $IDEF_{1X}$

    (c) $IDEF_{0/1X}$ Parser

    (d) Functional Interaction Configurator (refer to Section 5.3 for details).

It must be noted that (c) and (d) are tools which have been specially developed for integrated life-cycle support.

## 6.2.1 Information Model Enactment

The $IDEF_{1X}$ entity-attribute relationship modelling tool was used to model the global schema representing common information used by interoperating functional modules. It forms a composite view of a set of information models which corresponds to a generic reference model corresponding to the discrete parts manufacturing domain (and of the type described in Section 3.3.1). Thus the structure, content and entity relationships among the attributes of the information models are formally described. In this way the semantics are made explicit so that there is common interpretation of relationships among data items. The entity-attribute relationships of the information models is illustrated in Figure 6-2.

Typically, the model created contains data which is useful for database design and when specifying information requirement. Significant additional benefit can be realised by proceeding a step further, in terms of exploiting the data contained in the model, by semi-automating the implementation of database systems as defined by the information model. As previously mentioned, at present there is no generally agreed methodology or available supporting tools to realise model enactment in this way.

To overcome this limitation, a methodology has been conceptualised by the author, supported via a software toolset, by building on methods and tools previously developed by researchers at MSI in Loughborough, UK. The method will be described through reference to a proof-of-concept system devised to offer an insight into how model enactment can be effectively realised.

Here the EXPRESS data modelling language (which is an emerging standard used within ongoing STEP (Standard for the Exchange of Product data), initiatives world-

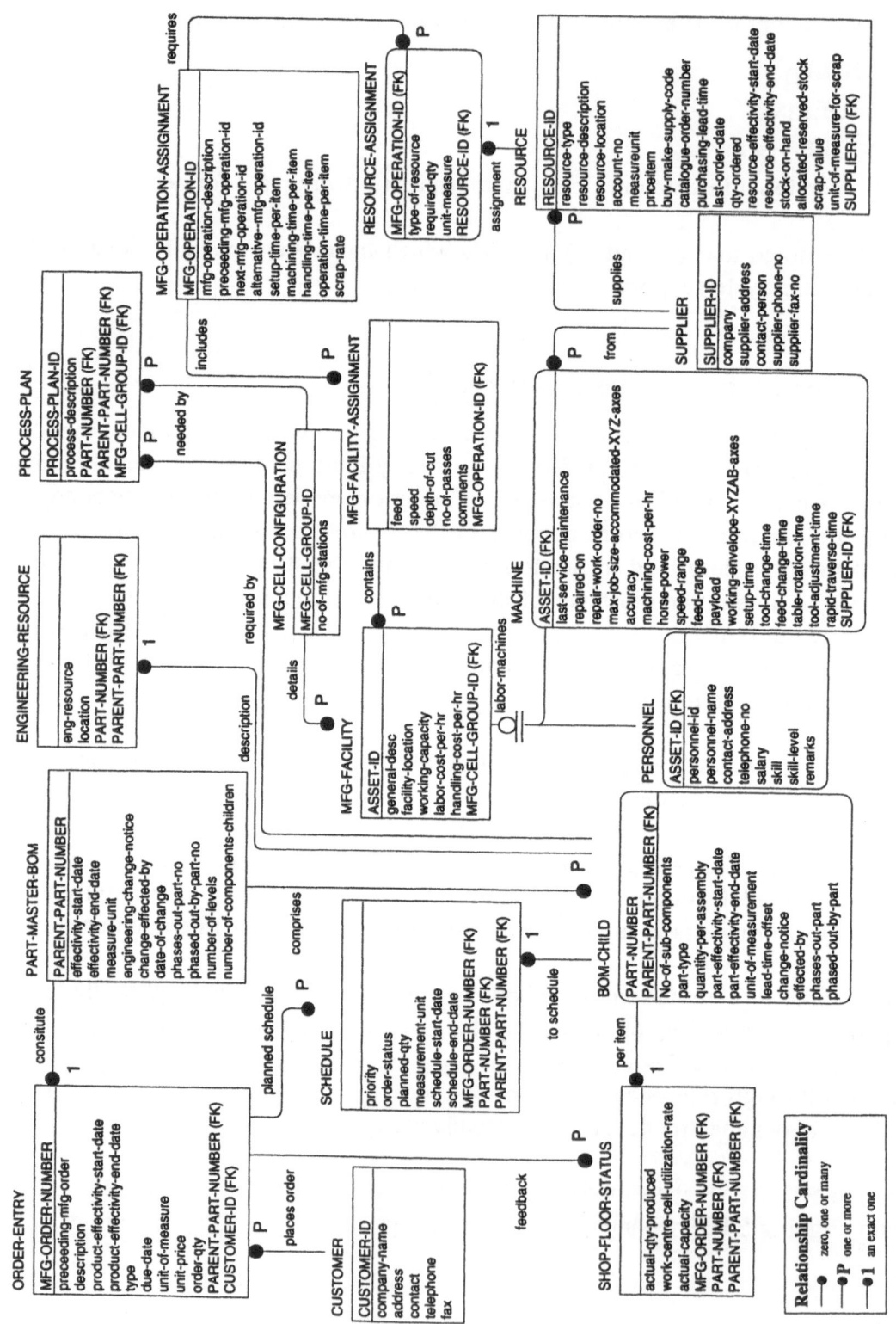

**Figure 6-2** IDEF1X entity-attribute relationship model representing a composite view of the information models

wide is used to facilitate detailed information modelling. EXPRESS is chosen as it enables implementation of a physical system in a manner which maintains the structure of the formal models.

**Figure 6-3** Set of build tools to develop data repository

As illustrated in Figure 6-3, the following tools were developed (Clements 1991a, b) to exploit EXPRESS as a means of representing aspects of the information architecture of CIM systems and providing a means of managing change more effectively:

- **EXPRESS to SQL Compiler** Using an EXPRESS to SQL Compiler software tool created by Clements at MSI, an EXPRESS information model can be directly compiled to generate SQL statements which will later be responsible for creating SQL compliant tables within the database. In this way the database is structured according to EXPRESS defined schemas so that relationships are strictly maintained; relevant information concerning the tables and their inter-relationships are automatically generated and stored in a dictionary. This approach is database independent (EXPRESS is not biased towards implementation) and can produce as output SQL statements which function for different relational database implementations, such as Ingres and ORACLE.

- **STEP Parser** This tool has been created to enable the population of the m tables (within different relational database) with real data in the format specified by the EXPRESS model. This process is structured by information contained in the output files generated by the EXPRESS to SQL complier.

An IDEF$_{1X}$ to EXPRESS transformation tool has also been developed by Clements at MSI to enable the automatic creation of an EXPRESS based information model from IDEF$_{1X}$ entity-attribute relationship descriptions. It operates by directly mapping the entities of the IDEF$_{1X}$ model into entities of the EXPRESS model. The establishment of this IDEF$_{1X}$ to EXPRESS computerised link not only offers a means of formally structuring information requirements but also supports the design, build and change processes associated with them.

The overall approach conceived by the author offers the following advantages:

• It helps simplify entity-attribute relationship modelling. This is possible because EXPRESS has a strong object-oriented nature where the information models are treated as objects, thus allowing modularity and data inheritance. It reduces the tedium and complexity associated with the schematisation of relational information models by eliminating the need to specify all common or primary attributes (which are necessary to link the tables together). Instead relationships among tables can be readily (a) established by allowing tables to inherit the attributes of other tables to which they are linked, and (b) altered and relationships added or removed without causing chaos in what remains.

• Allows easy extension and modification of information models.

• Supports and encourages the reusability of information objects, thereby reducing the development time and enabling wider scope solutions.

• Enables the use of IDEF$_{1X}$ created models, this being important in system development and maintenance phases of projects.

• Implementation is database independent.

## 6.2.2 Functional Activity-based Modelling

The process of modelling functional activities is generally concerned with broad issues related to the formal identification and description of functional properties and dependencies; not just on specific details to the required capabilities of each functional unit.

Any activity-based modelling tool which conforms to the $IDEF_0$ methodology can be used to produce a functional activity-based model. It can define dependencies and inter-relationships among the interoperating functional modules in a structured way. Successive decomposition of such an activity-based model (in a hierarchical manner) can be realised to define interactions among the functional activities in greater details; where inputs required to drive the functions and their outputs (generated under supervision of their controls and enabling mechanisms) can be represented. For example, a hierarchical decomposition of a functional activity-based model modelling interoperation among production planning, finite capacity scheduling, process planning and SFCM activities is illustrated in Figures 6-4, 6-5 and 6-6.

## 6.2.3 System Behaviour Enactment

A formal definition of the interaction processes between functional components of an integrated system requires complete and accurate descriptions of (a) the flow of information between function blocks, and (b) the form and type of information; this needs to be made available to support and drive those functions, so that they can realise their assigned tasks. As previously discussed there should be a means of:

- unifying the perspectives of function and information modelling in order to establish an association between the two modelling streams; and

- facilitating the structuring of downstream life-cycle processes, for example to aid implementation and configuration in relation to co-ordination and control for functional interaction.

One novel approach to meet these requirements involves the use of an $IDEF_{0/1X}$ Parser tool and the configurator which together can formally define and describe the behavioural aspects of an interoperating system where the

- $IDEF_{0/1X}$ Parser is responsible for the enactment of functional activity-based and information models in order to establish association between them; and

- the configurator generates data which encodes the required associations (refer to Section 5.3 for further details on the configurator).

The $IDEF_{0/1X}$ Parser and configurator tools need to operate in close harmony with each other. All relevant data concerning the functions, information entities and the nature of their associations (i.e. as either input or output) are required to be stored in a shared database (e.g. the functional interaction module database). These data

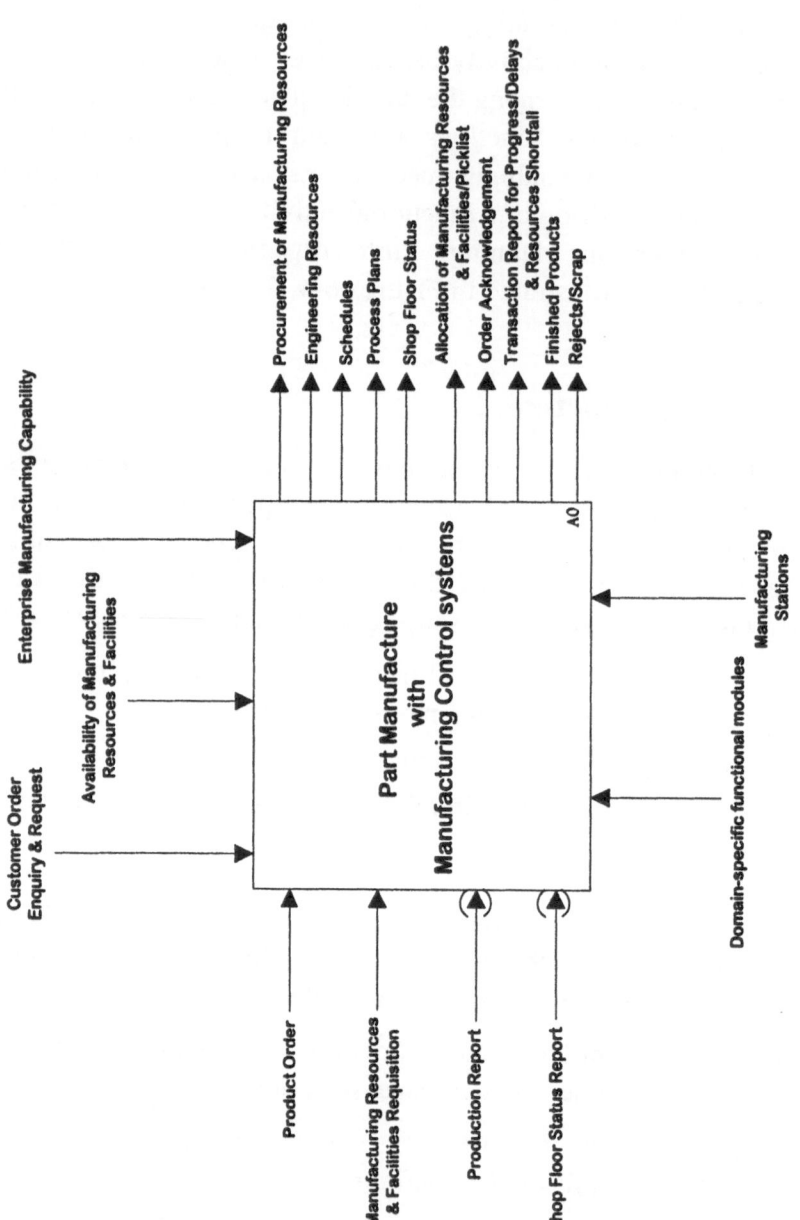

**Figure 6-4** IDEF$_0$ function model — context diagram

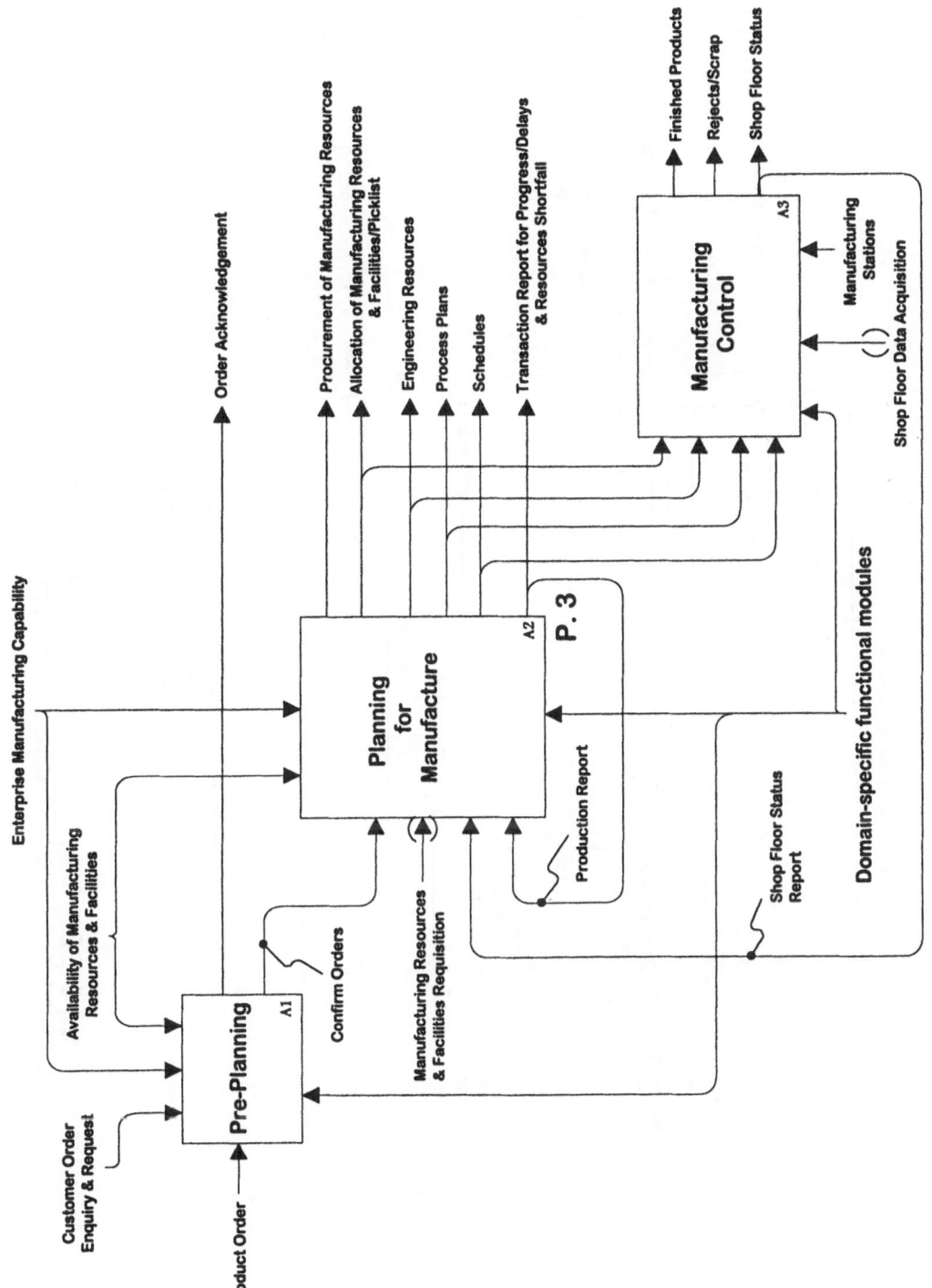

**Figure 6-5** IDEF$_0$ function model to describe production planning
and its inter-relationship with other functional modules

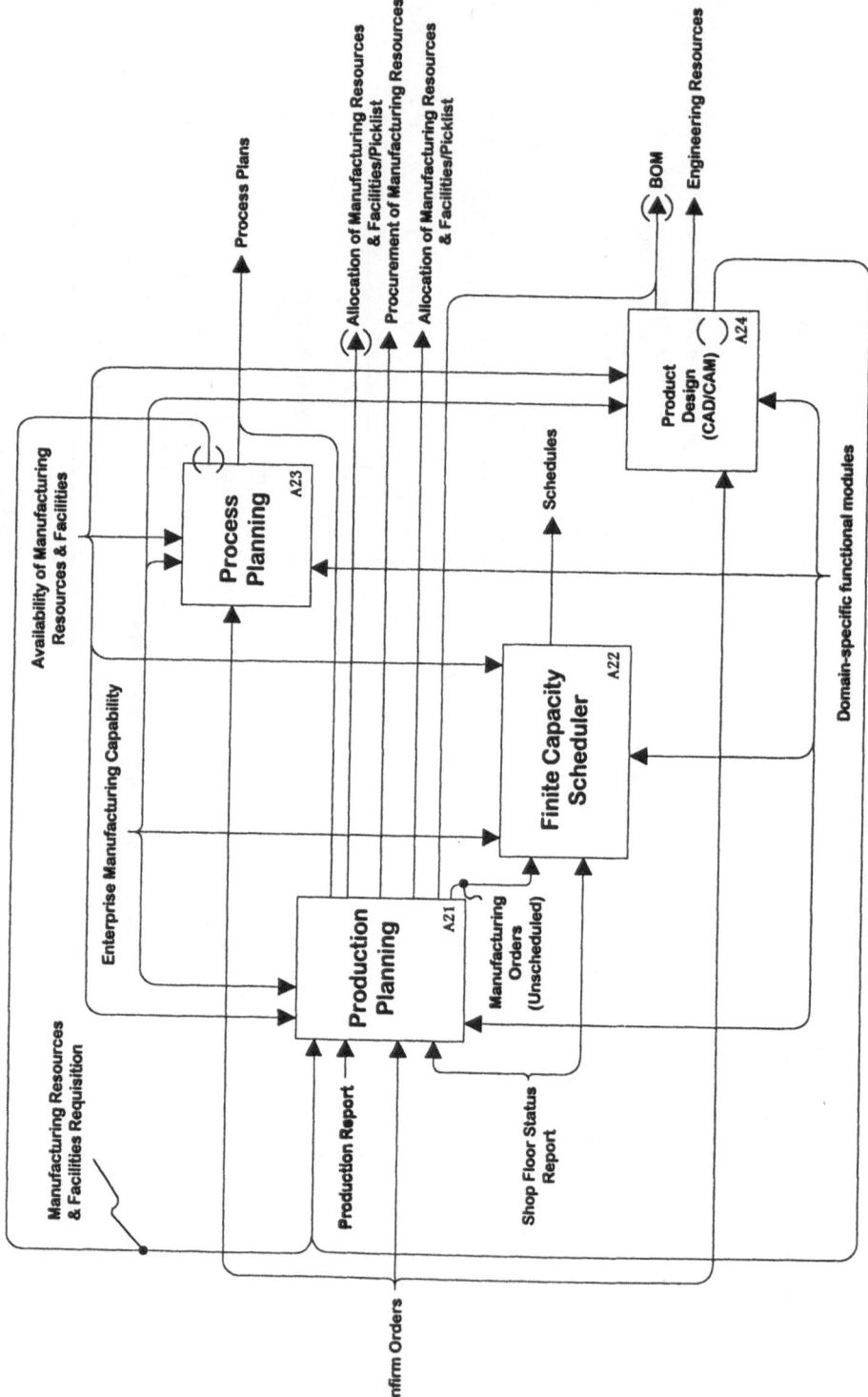

**Figure 6-6** IDEF₀ function model with detailed level of decompositon for planning for manufacture activity within production planning domain

effectively serve as working parameters which characterise particular properties of the system during normal operation.

As illustrated in Figure 6-7, the tools perform the following specific tasks:

- **IDEF$_{0/1X}$ Parser** This sorts through and assimilates large amounts of data generated in the IDEF$_0$ and IDEF$_{1X}$ reports which contain details pertaining to the models. It is primarily responsible for the following:

    - selection of required functions defined in the IDEF$_0$ activity-based model and information entities defined with the IDEF$_{1X}$ tool; and

    - population of the selected data into a working database.

- **Configurator** The configurator processes the data derived from the IDEF$_{0/1X}$ Parser. It offers services which enable the user to interactively set, edit and display associations between functions and information. The configured associations between functions and information entities are stored in the function-information association table and used (during run-time) for the following purposes:

    - to ensure accountability of information to functions; and

    - to control and co-ordinate the functional work flow in manner where information availability and completeness are verified prior to enabl. any function requiring access to it.

Figure 6-7 illustrates the application of the IDEF$_{0/1X}$ Parser and configurator to the IDEF$_0$ activity-based and the IDEF$_{1X}$ entity-attribute relationship models (as exemplified in Figures 6-2, 6-4, 6-5 and 6-6) by processing their generated reports. This helps to highlight in a concise manner the ability to enable and support the establishment, management and change of associations between functions and information.

The methodology based on the use of IDEF$_{0/1X}$ Parser and configurator tools offers an effective and formal means of capturing and establishing a link between functional components and their required information. No extra effort is placed on the system designer or builder, except for configuring the required associations based on data drawn directly from existing function (or activity-based) and information models. In addition, the consistency and accuracy of the associations between function and information is ensured and maintained from a very early stage in the system life-cycle

**Figure 6-7** Methodology for function-information association

(i.e. the conceptual design stage) right through to system implementation and maintenance. This reduces uncertainty and resolves many potential conflicts. In addition, this approach also helps improve the accountability and traceability of functions and information.

## 6.3 SUMMARY

As illustrated in this chapter, system design and modelling tools can be deployed to provide an effective entry point to support the life-cycle of systems in an integrated manner. In this way results and knowledge generated at one phase can be used during other life-cycle phases. Models generated using one or more chosen modelling method can provide a formal representation of different views (i.e. function, information, behaviour) of the system under consideration and can serve as a source of usable data to structure and enable various downstream life-cycle processes.

However, in order to truly achieve and enable integrated life-cycle support it is necessary to allow usable data to be referenced, accessed, manipulated and formatted in a manner suitable for use during each life-cycle phase. However, as clearly highlighted from the case study reported in this chapter a useful first step towards meeting this requirement is to identify and make available the necessary software toolset, such as the following, in order to bridge existing gaps in capability:

- EXPRESS to SQL compiler
- STEP Parser
- $IDEF_{1X}$ to EXPRESS transformation tool
- $IDEF_{0/1X}$ Parser
- configurator.

Each of these tools exists as an independent entity but in the author's research study were linked through a shared working database where usable data is stored to enable common access and usage, and thereby support the different life-cycle phases. It should be noted that the software toolset developed can be configured to support other system design and modelling methodologies on condition that the methodologies in question provide ready access to their underlying data and knowledge structures which they use to encapsulate function and information models.

Finally, in relation to the overall methodology described, which enables interoperation between functional components (see Section 3.5), this chapter has reported on two sub-methods, namely:

- means of enacting function models
- means of enacting information models

in a manner which can realise synergy between legacy software previously designed to operate in a stand-alone way.

In view of a generalised need to adapt and respond to changes in manufacturing system requirements, the set of build tools described herein can provide the system builder with a formalised and structured approach to the creation and maintenance of integrated manufacturing systems.

# 7 Use and Appraisal of the Methodology Derived

## 7.1 INTRODUCTION

In order to illustrate the application and ascertain the level of effectiveness of the methodology derived, which seeks to enable interoperability among functional components, a proof-of-concept implementation study will be presented and discussed in this chapter. This study constitutes a part of the author's research work. The implementation study is described in a concise and general manner to exemplify the overall methodology.

In the implementation study a proof-of-concept system was built to demonstrate the interoperation of a number of typical software applications which are representative of the manufacturing control domain. The main element of the system is a commercially available stand-alone proprietary MRP II based CAPM package, namely Manufacturing Control Code (MCC). The methodology, and its tools and infrastructure, transforms this package into a more open system so that it can more readily interoperate with other functional components. This demonstrates a particularly important industrial capability, namely the integration of existing legacy (or 'as is') software applications. The following sub-systems were utilised:

- functional interaction module which embodies the services required to facilitate domain specific functional interaction and to govern system behaviour (during run-time);

- CIM-BIOSYS IIS to simplify interconnection and to provide common general purpose integration services which support functional interaction; and

- set of build tools to formally structure (a) interaction among functional components, and (b) implement and maintain information system on a system-wide basis, by managing and maintaining a data repository which holds information of common concern.

*The CIM Debacle*

**Figure 7-1** Illustration of proof-of-concept system for software interoperability

## 7.2 PROOF-OF-CONCEPT IMPLEMENTATION

As illustrated in Figure 7-1, the following functional components constitute the complete proof-of-concept system:

- **Production Planning** This encompasses order entry, scheduling, and manufacturing resource and facility management which includes allocation, procurement as well as routing for part manufacture. MCC, which is a commercially available CAPM package from John Brown Systems PLC (UK), was chosen for this study as its information is stored in a relational database, namely ORACLE RDBMS, and is to a certain degree accessible by other applications. The purpose of MCC is to turn product orders into manufacturing schedules for enaction by a set of manufacturing resources (refer Figure 7-2).

- **Finite Capacity Scheduler** Although the finite capacity scheduling function available in MCC is utilised, it is intentionally treated as an independent and separate function in order to isolate and effectively study its interaction needs with other functional components. This finite capacity scheduling sub-system is responsible for the short term planning of manufacturing orders in a manner which optimises manufacturing operations on the shop floor. It relies on the availability of data from production planning, treating it as its input, to perform the necessary scheduling function based on, for example, resource and manufacturing facility allocation and constraints, part procurement and manufacture lead time as well as routes for part manufacture.

- **Shop Floor Control and Monitoring** A SFCM application software was integrated into the proof-of-concept system. Here it is required to despatch planned manufacturing orders (scheduled by the finite capacity scheduler) and to co-ordinate, execute, control and monitor the operation of shop floor activities. It is also responsible for shop floor data acquisition thereby enabling and generating production status feedback.

- **Decision Support System** This class of software system was represented by a dynamic cost analysis tool which was created on a need basis to support strategic decision-making for economic part manufacture (Shaharoun *et al.* 1992). This decision support system has been included in the proof-of-concept system in order to highlight the ease with which the system can be

**Figure 7-2 MCC's information requirement and flow**

expanded in terms of building and plugging in *ad hoc* specialised functional components that require access to data available in the system data repository. It relies on both planned and actual production data, derived from production planning and shop floor control and monitoring respectively, to perform the necessary cost analysis.

The system was built to facilitate interoperation between the above-mentioned functional components in a manner to enable sharing of and access to information of common interest (see Figure 7-3 which depicts information flow between functional components and dependency between information entities).

**Figure 7-3** Overview of information flow and dependency
between information entities

## 7.2.1 Implementation Steps

The following four inter-related meta-steps were supported by the author's methodology in a way which structured the development of the proof-of-concept system:

- implementation of a system-wide information repository;
- flexible interconnection of the selected functional components;
- establishment and configuration of appropriate functional interaction capability; and
- establishment and configuration of user interface capabilities.

Details of the activities carried out in each of these meta-steps are outlined as follows:

### Implementation of a System-wide Information Repository

The following representative example activities are typically involved in establishing and managing the manufacturing information system:

**(a) Identification and formal representation of information requirements**
A global schema corresponding to system-wide shared information, i.e. shared between the functional components concerned, was represented in the proof-of-concept system with reference to the generic information models. It represents a unification of information entities of common concern (to be held in the system data repository) which will require to be accessed and updated. Here $IDEF_{1X}$ was used to formally represent the global schema (as illustrated in Figure 7-4). It uncovers semantic properties of the underlying information system and makes them explicit within the data definition of information objects.

**(b) Mapping between local database to the system data repository**
Proprietary information, stored in the MCC relational database, which is of common interest to other functional components, was mapped onto the system data repository via reference to the global schema (refer to Figures 7-5, 7-6 and 7-7). This same mapping principle can be applied for all legacy components, in which proprietary information data structure is used, thereby enabling other system components to gain access and update that information in a flexible manner.

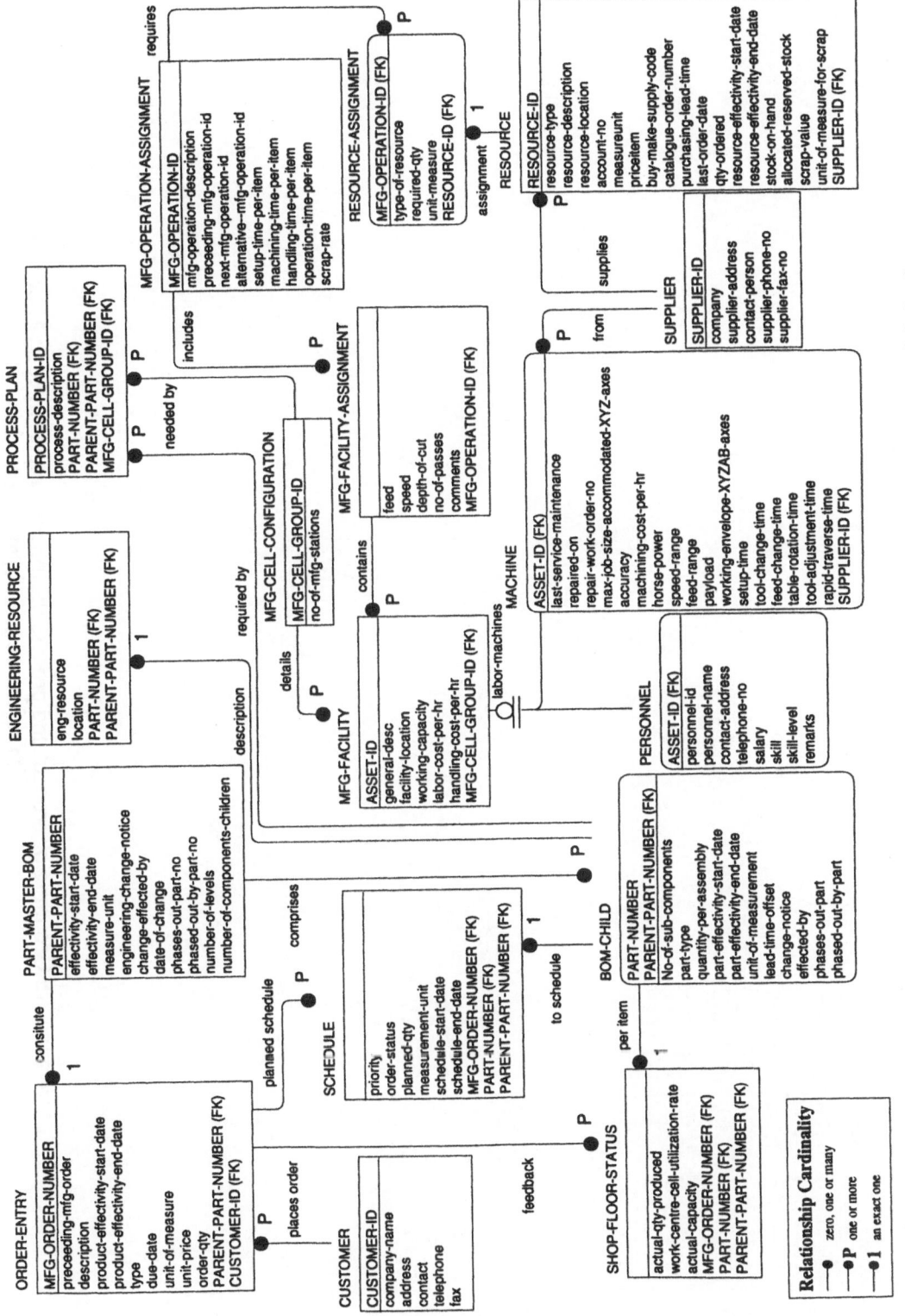

**Figure 7-4** IDEF1X entity-attribute relationship model representing the global schema

**Figure 7-5** Grouping of product and process information represented in MCC database with reference to information models

**Figure 7-6** Grouping of schedule information represented in MCC database with reference to information models

**Figure 7-7** Mapping between information entities

## (c) Creation of relational tables stored in the system data repository

The set of build tools used here to develop the system data repository and the corresponding data generated and processed by them at each stage are depicted in Figures 7-8 and 7-9 respectively.

First the $IDEF_{1X}$ to EXPRESS transformation tool is applied to automatically create an EXPRESS based information model which has an underlying data structure related to the entity relationships defined in the global schema, i.e. during activity (a) by using the $IDEF_{1X}$ modelling tool. The reader should refer to Appendix IV for details on the

**Figure 7-8** Set of build tools to develop data repository

EXPRESS based information model used as the global schema of the proof-of-concept system.

The EXPRESS information model is then compiled by applying the EXPRESS to SQL Compiler to generate SQL statements (responsible for creating relational tables which will be stored in the system data repository). In this case, the system data repository was chosen to be ORACLE based. ORACLE has a similar nature to that of the MCC database, and this simplified to some extent, information sharing and transfer between the two database. Hence the system data repository was structured according to the EXPRESS defined schemas which has the important benefit that relationships are strictly

**Figure 7-9** Information flow to support the development of the data repository

maintained, resulting in relevant information contained in the tables and inter-relationships among information objects being automatically generated and stored in a data dictionary via use of the STEP Parser.

### (d) Populating the system data repository

At run-time the shared information (stored in the system data repository) is accessed through using services offered by a database 'driver', thereby making them accessible to the other functional components. Thus the system data repository effectively serves as the focal point for exchanging, updating and sharing information of common concern between the various functional components concerned. The services of the database 'driver' and the data access objects are required to have a knowledge of the mapping processes (defined during activity (b)) between local and global database schema as illustrated in Figure 7-7.

Using a similar principle, the data store of the SFCM application was mapped onto the system data repository; this was also to establish a share information capability within the proof-of-concept system.

### Flexible Interconnection of the Functional Components

The task here is to interconnect the various functional components which will normally be heterogeneous in nature and reside at different computer nodes of a distributed system. As previously discussed, in order to achieve flexible interconnection among such components, an IIS would be required. Here the CIM-BIOSYS IIS was chosen. However, viewed from the perspective of a CIM-BIOSYS system builder MCC is a non-conformant (or alien) application, i.e. it is not inherently compatible with the CIM-BIOSYS IIS architecture. Thus it was necessary to design and implement an 'alien application shell' specially developed for MCC to provide it with sufficient capability to use the common integration services offered by the CIM-BIOSYS IIS (which includes inter-process communication and information management). Conversely, the SFCM and the decision support systems are conformant applications and they can operate directly over the CIM-BIOSYS IIS; hence they have no need of an 'alien application shell'.

The functional components and the system data repository need to be registered users of the CIM-BIOSYS IIS. This registration requires their actual physical location over the local area network to be clearly specified and stored within a CIM-BIOSYS configuration file. This knowledge is essential to the CIM-BIOSYS IIS during its normal operation in order for it to (a) activate required functional component; and (b) enable independent and transparent information access via the database 'driver', i.e. without the need for users to have detailed knowledge of integration issues.

### Establishment and Configuration of Functional Interaction Capability

In this meta-step, as illustrated in Figure 7-10, the $IDEF_{0/1X}$ Parser is used to process the activity-based functional model generated via $IDEF_0$ (as described in Figures 6-4, 6-5 and 6-6 in Section 6.2.2) and the $IDEF_{1X}$ entity-attribute relationship model (described in Figure 7-4) as encoded by their generated reports. This process can be viewed as enacting the function and information models in order to establish associations between them. The reader should refer to Figure 6-7 in Section 6.2.3 for an illustration of principles involved here.

Subsequently the configurator is used to formally establish and configure associations between the functional activities and their shared information requirements, represented by the function-information association table of the functional interaction module. The table is referenced by the interaction management service (during run-time) to govern the dynamic behaviour of the system in a controlled and co-ordinated manner based on predefined sequences of activities and information needs established between functional activities.

**Figure 7-10** Configuration of functional interaction module

**Figure 7-11** 'Application shell' configured to coalesce functional components and integrate with functional interaction services and CIM-BIOSYS IIS

**Establishment and Configuration of User Interface Capabilities**

As illustrated in Figure 7-11 and earlier described, a configurable 'application shell' was created to act as a generic front-end user interface. In the proof-of-concept system this 'application shell' was configured to:

- provide a consistent and simple end user interface to the various functional components, thereby providing an effective working environment for human interaction during run-time; and

- provide end user access to software toolset, CIM-BIOSYS IIS and functional components, in a way which facilitates system construction, management and change.

## 7.2.2 Analysis and Discussion

The following section summarises the most important observations and findings resulting from the implementation and running of the proof-of-concept integrated and interoperable system:

### Development Process

(a) A higher entry point to integrated system development is offered which provides a structured path towards a more organised and prescribed approach. This property is attributed to the application of generic reference and activity-based functional models which capture and encapsulate generic working knowledge of part manufacture in terms of clearly specified and defined information flows as well as relationships between functional activities and their information needs. This approach avoids continuously re-inventing the wheel, which is a common disadvantage of developing custom built solutions from scratch.

(b) It is necessary to have detailed knowledge of the proprietary information structures used in proprietary database (MCC in this case) in order to understand their semantics and structure sufficiently well enough to share information in a flexible and effective manner. Such knowledge is required before the relevant information can be accessed or mapped from the database of any 'as is' business application onto the system data repository, thereby establishing its common usage. Thus it is vital to gain a sufficient level of support and understanding of would be interoperating software products from their manufacturer (products provided by John Brown Systems PLC for MCC in this study). Without such knowledge there will be severe restraints on further progress in the development of interoperable systems.

(c) When incorporating 'as is' business applications (of the MCC ilk) into interoperating systems, some degree of data duplication is inevitable. However, as an underlying axiom of the methodology adopted by the author is to decouple functional activities from their information repositories, local changes (i.e. on the locality of a single component) to information and functional aspects of an application component will have minimal effect on other applications. Thus change management is much facilitated and it can be expected that functional and information needs of the system can be handled in a largely separate and less complicated manner.

(d) The system data repository can be expected to be only partially populated (such as with information of common concern to a single data store, like the MCC database); thus a degree of data independence is offered where the individual local databases retain their autonomy and hence can continue to serve their existing customer set.

(e) The set of build tools used in this case study demonstrated how a more systematic and structured approach can be provided to engineering interoperable and integrated manufacturing systems solutions. The software toolset enables systems life-cycle support in a transparent and co-ordinated manner from design to implementation so as to facilitate ease of system development and change management.

## Operation During Run-time

(a) The functional activities (contained within interoperating components) are made available to end users through the configured 'application shell', which coalesces and integrates the interworking of the functional activities via the tools and services provided by the functional interaction module and the CIM-BIOSYS IIS. This can result in considerable synergy, not only in providing any single end user with access to specialised functional activities to support part manufacture but also in enabling the sharing of concurrent knowledge in a way which can support better and more informed decision-making. For example, the end user is able to take into account information regarding availability of manufacturing resources and facilities as well as manufacturing capability when considering the requirements and specifications of parts accepted for manufacture. This is particularly useful for the inclusion of legacy systems where they can be brought to an acceptable level of conformance to utilise the services offered by the IIS via the 'application shell'. Arguably, behavioural codes for conformant business applications can be generated and embedded within them so as to enable their direct interoperation via the IIS. However, for ease of management and change they should be made available in the form of APIs where they can be readily incorporated when required rather than be inextricably linked to the application software functional codes.

The proof-of-concept system has been shown to promote intra-organisation integration, where an effective and defined channel for dissemination of knowledge and information is made available via a consistent and effective front-end user interface.

(b) Discipline is well maintained when functional components interact during mm run-time. As functional interaction is based on data availability and clearly predefined sequences of activities, any conflicts or misunderstandings might occur, in relation to information flows and needs, are effectively eliminated. For example, when the production schedule is required by the SFCM system to co-ordinate and govern shop floor execution, the end user issues a request through the configured 'application shell' for access of the production schedule (stored in the system data repository). The interaction management service of the functional interaction module would interpret the end user's request and check for availability of the production schedule via its status management mechanism. The interaction management service also ensures that the production schedule has been processed and prepared by the finite capacity scheduler in accordance with the sequence of activities predefined in the Data I/O Table, prior to triggering off the SFCM system to proceed with access of the requested information from the system data repository. Hence co-operation among interoperating software components is enabled which is not normally possible by conventional means.

Finally, it should be highlighted that the proof-of-concept system described here offers an effective means of improving interworking among finite capacity scheduling, SFCM and production planning systems, and represents a step towards more open versions of such systems which could provide even greater benefit from such a facility. This in turn will help to narrow the wide gap which currently exist between upstream production planning and downstream shop floor activities.

# 8 Conclusion

## 8.1 GAINING A BROAD PERSPECTIVE

Software interoperability is a major sub-goal of systems integration which is aimed at facilitating physical and business application integration. As discussed, there are many important issues which need to be carefully scrutinised in order to effectively realise an integrated solution. The underlying principle is to understand and rationalise intrinsic business requirements and to identify and exploit appropriate enabling technologies to help enhance productivity and efficiency on an enterprise-wide basis. However, it is imperative that this should not be limited by capabilities and performance characteristics of available technology because businesses would inevitably evolve to suit changing needs and there would also be progressive advances made in technology.

The author strongly advocates the need to (a) adopt a more formal and structured approach to the design, development, support and enhancement of integrated systems through their life-cycle, and (b) simplify interconnections so as to facilitate functional interaction among the various software components in a concurrent and coherent manner where information of common interest can be shared in a highly flexible and standardised manner. To enable software interoperability the following requirements (as identified in Section 1.4) need to be facilitated:

- information sharing
- interconnection
- control of system behaviour
- system design and development

### Information Sharing

- Generic reference models which constitute prime information required by the various functional components underpin the sharing of information of common interest (such as those identified and specified by the author for production planning, product design, finite capacity scheduling and SFCM systems). They are characterised by their generalized applicability, while being sufficiently flexible to enable customisation to suit specific user needs. Indeed the reference models offer promise as being effective in addressing current problems which result from a lack of standardisation in information representation and exchange for functional components.

- An information architecture needs to be incorporated which establishes structure and uniformity whilst enabling sharing and transfer of information among functional components via use of the generic reference models. The information architecture acts as a global library of information entities, providing mechanisms for representing and managing physical data; which in a typical system is actually stored in a fragmented fashion within a number of heterogeneous data stores. Hence the information architecture provides a foundation for defining, identifying, and integrating both specific and generic information entities.

## Interconnection

- The IIS is used to structure and simplify problems of realising interconnection among functional components. It separates integration and application issues in a manner which resolves differences in the physical system relating to heterogeneity, distribution and data fragmentation.

Generally, the IIS provides a combination of low and high level common integration services which help realise physical integration processes. Inter-process communication and information management and access constitute part of the low level services where, for example, it ensures consistent, reliable, transparent and open access of information stored in the data repository (via its available 'drivers') thus making it available to distributed processes in a device independent manner. It maps distributed processes (embodied in functional components) onto the physical data repository contained within a target manufacturing system. The high level services are aimed at offering 'soft' or flexible integration of functional activities to enable their reconfiguration and incremental development.

## Control of System Behaviour

- Capability to facilitate functional interaction is essential to (a) formally and flexibly structure threads of functionality embedded within various functional components, and (b) facilitate their interaction (during run-time) in a controlled, co-ordinated and deterministic manner, where considerations for associated preceding and succeeding activities and for their shared information needs are taken into account.

It offers an effective framework for governing system behaviour in an event- and data-driven manner based on functional dependencies and information needs and availability. The functional interaction management services, as discussed in Chapter 5, can be viewed as constituting high level integration mechanisms and tools of an IIS. It builds upon the low level, general purpose integration services and tools of the IIS, which provide standard data inter-communication and information transfer facilities to interconnect software components.

There is also a need (via an appropriate front-end user interface) to ensure consistent access to and coalesce the interoperation of various functional activities over the IIS which:

- enables access to the various functional activities which may be physically distributed so as to provide the end users with a system-wide working viewpoint to effect better and more informed decision-making in relation to the specific tasks they perform;

- makes functional interaction management easier by effectively insulating the end user from the complexities and tedium involved, which will be taken care of by the relevant IIS and functional interaction management services offered.

## System Design and Development

- Many system design, modelling and analysis methods are available and they vary in nature depending upon the important views they represent such as organisation, process, control, behaviour, resource, cost, information, functional activities, etc. The purpose is to model the physical environment under consideration as closely as possible and to represent it in a manner where it can be well understood. This allows for more informed decision-making when performing analysis.

However, as pointed out there is no one single system design, modelling and analysis tool available today to handle all of the various life-cycle phases or various views or levels of model genericity. Thus the model-driven integrative approach is advocated by the author. It uses models to formally structure and support implementation, run-time and change processes in a way which supports the various life-cycle phases of systems. Here the information and activity-based models created are exploited during downstream life-cycle phases to ensure clarity, consistency, accuracy and re-utilisation of knowledge and data between phases. In order to enable this a set of life-cycle support tools is need (such as those discussed in Chapter 6 which couple closely with the $IDEF_0$, $IDEF_{1X}$ and EXPRESS-based modelling tools used for activity-based, entity-attribute relationship and data modelling respectively).

Accordingly, these tools reference, access, manipulate and reformat data (corresponding to the functions and information models) so that they assume a form which is suitable to structure and enables various downstream life-cycle processes, as discussed in the previous chapter with reference to the proof-of-concept study that has been carried out. Hence the software toolset offers system builders a more formalised and structured way of creating and maintaining integrated manufacturing systems, thereby catering for their need to adapt and respond to changes in system requirements. This enables overall system reconfigurability, more optimal system design and operation and a reduction in the time and effort involved in creating such systems.

The realisation of a universally accepted 'fully specified open standard for software interoperability', which enables on a widespread basis unconstrained interaction and interchange between heterogeneous software components, would of course be much desired. However, it is currently an impractical goal due to the enormous complexity of the problem and to the many outstanding standardisation issues which have yet to be resolved. Much of it depends upon:

- availability of suitable functional modules with facilities required for the following which should be sufficiently standard to make many people adopt and write to those standards:
  - inter-process communication
  - information sharing and management
  - interconnection;

- availability of acceptable reference models which can describe in a comprehensive manner (a) information flow and requirement, (b) functional activities and their inter-relationships and dependencies; and (c) system behaviour;

- availability of life-cycle support systems to design better integrated and interoperable solutions; and

- progressive release of more interoperable functional components (with more modular and atomic functionality) from vendors.

However, on the other hand, the use of contemporary 'turnkey' or 'custom built' solutions (which do not adhere to any open standards) is not tenable for systems requiring interoperation of many functional components. Constraints in software interoperability will undoubtedly remain if such proprietary systems are not designed to enable access to their threads of functionality or underlying information.

Bearing in mind these difficulties, the author has described a realistic approach which can be applied today to realise improved synergy between contemporary software packages used to support the manufacture of discrete products. As such the approach can be considered to be part-way between the extremes of 'open' and 'closed' systems, as illustrated in Figure 8-1.

**Figure 8-1** Degree of software interoperability

The emphasis has been to provide a means of enabling a degree of software interoperability which overcomes (a) limitations inherent in contemporary functional components and solutions, and (b) associated and inherited problems concerned with achieving their interoperation. Hence the aim is to offer an infrastructure that enables a degree of interoperation which not only allows for the adoption of legacy (or 'as is') functional components but also the introduction of a new generation of highly reconfigurable, modular and more open software products, i.e. 'to be' products (as they become available to industry). A migration path is offered towards 'open' manufacturing systems which are more readily adaptable in the face of changing functional and operational requirements.

In order to provide for more effective and universally applicable interoperable solutions, the author expects future interoperability development in the following major areas:

- advancement of life-cycle support approaches leading to simulation, emulation and execution of functional components and their integration into custom designed systems so as to result in better designed systems;

- standardisation of information reference models and activity-based functional models;

- improved and standard infrastructural facilities;

- progressive development of a new generation of more open functional components (possibly on the base of general software developments).

# Appendix I

## INFORMATION MODELS

| |
|---|
| Manufacturing Facility |
| Part Master/ BOM |
| Resource |
| Process Plan |
| Order Entry |
| Schedule |
| WIP |
| Engineering Resource |
| Manufacturing Cell |
| Customer |
| Supplier |

Schedule

| FIELD DEFINITION | SIZE | DESCRIPTION |
|---|---|---|
| Manufacturing order number | N[7] | User defined unique identifier for a batch of manufacturing order. |
| Part number | C[15] | Unique identifier for the part (or component). |
| Priority | N[3] | It sorts and dictates the sequencing of scheduled jobs. |
| Order Status | C[1] | It indicates the part's current position within the order process. |
| Schedule Status | C[10] | It indicates the status of the part within the scheduling and manufacturing process. |
| Planned Quantity | N[9] | It specifies how many or how much of the part is required for manufacture. |
| Unit of Measure | C[4] | It is the standard quantitive unit for the part used in the manufacturing process. |
| Schedule Start Date | Date | Planned date of commencement for the manufacture of the required quantity of parts. |
| Schedule End Date | Date | Planned date of completion for the manufacture of the required quantity of parts. |

(Schedule / Unschedule / Complete / WIP / Hold)

2 = Quotation or Firmed planned forecast

3 = Open Order. Confirmed order but all required issues or shipment have not been made for the item.

4 = Released Order. Confirmed and planned but pending receipt of required issues or shipments.

5 = Closed Order. All required issues or shipment have been made for the item and can therefore proceed with manufacture.

6 = Closed Order. Manufacture of the item is complete and is ready for dispatch.

7 = Closed Order. The order is ready to be deleted from the active file and retained in order history.

8 = Closed Order. Purge the order and do not retain in order history.

9 = Credit Hold. The customer's credit limit has been exceeded or the order is placed on hold for another reason. The item is then treated as an open order.

**Part Master/BOM**

The bill of materials (BOM) is a list of the items, ingredients or materials needed to produce a parent item, end item, or product. It is not just a simple listing of dependent demand items, but a structured list which describes the sequence of steps in manufacturing the product.

The BOM is utilised for the following:

* Manufacturing engineering use the BOM to show how to manufacture the product.
* Production planning use the BOM to schedule the parts which make the product.
* Manufacturing use the BOM to make the product on the shop floor.
* Finance use the BOM to cost the product.
* Order entry use the BOM to translate customer orders and enquiries based upon the inter-dependency and relationships of components for the product in terms of manufacture, procurement and etc.

| | FIELD DEFINITION | SIZE | DESCRIPTION |
|---|---|---|---|
| **Part Master/Parent** | Parent Part Number | C[15] | Unique identifier for the parent item which is the higher level item in the BOM. |
| | Effectivity Start Date | Date | Date on which the part is introduced into the BOM. |
| | Effectivity End Date | Date | Date on which the part is severed from the BOM |
| | Unit of Measure | C[4] | It is the standard quantitive unit for the part used in the manufacturing process. |
| | Engineering Change Notice Number | N[7] | The revision for the BOM as authorised by the engineering department. |
| | Change Effected by | C[20] | Personnel, section or department responsible for the change. |
| | Date | Date | Date on which the engineering revision was initiated. |
| | Phases out Part Number | C[15] | Preceeding part that was in use. |
| | Phased out by Part Number | C[15] | Succeeding part to be used. |
| | Number of levels | N[2] | Number of levels in BOM supported for the parent item to be produced. |
| | Number of children | N[3] | Number of different sub-components supported at the specified level. |
| **BOM Child** | Parent Part Number | C[15] | Unique identifier for the parent item which is the higher level item in the BOM. |
| | Number of Children | N[3] | Number of different sub-components supported at the specified level. |
| | Part Number (Child) | C[15] | Unique identifier for the part (or component). |
| | Part Type | C[1] | It distinguishes the various types of relationship between a component and its parent item in the BOM. |
| | Quantity per Assembly | N[12,2] | Number of components required for assembly of per unit parent item. |
| | Effectivity Start Date | Date | Date on which the part is introduced into the BOM. |
| | Effectivity End Date | Date | Date on which the part is severed from the BOM |
| | Unit of Measure | C[4] | It is the standard quantitive unit for the part used in the manufacturing process. |
| | Lead-time Offset | N[9] | It is the difference between the due date and the release date. |
| | Engineering Change Notice Number | N[7] | The revision for the BOM as authorised by the engineering department. |
| | Change Effected by | C[20] | Personnel, section or department responsible for the change. |
| | Date | Date | Date on which the engineering revision was initiated. |
| | Phases out Part Number | C[15] | Preceeding part that was in use. |
| | Phased out by Part Number | C[15] | Succeeding part to be used. |

N = Normal component that is consumed in the manufacture of its parent.

P = Phantom component that is used for BOM structuring purposes only (e.g. a transient subassembly consumed in the manufacture of its parent).

R = Resource component used in the planning process of the manufacture of its parent (e.g. labour and machining hours).

C = Co-product component derived from the manufacture of the parent.

T = Tool component used in the manufacture of the part.

U = Tool return item which will be returned after manufacture of the part.

**Process Plan**

The process plan will involve assignment of manufacturing facilities and resources to each of the planned manufacturing operation. The process routing is geared towards cellular type of production where Group Technology is applied to identify 'sameness' of parts, equipments or processes in order to derive suitable working cells.

| FIELD DEFINITION | SIZE | DESCRIPTION |
|---|---|---|
| **Process Plan Identification** | | |
| Part Number | C[15] | Unique identifier for the part (or component). |
| Process Plan ID | N[7] | Unique identifier for the process plan. |
| Process Description | C[60] | Brief textual description pertaining to the process routing. |
| Manufacturing Cell Assignment | N[3] | Unique identifier for assigned manufacturing cell where part need to be delivered for manufacture. |
| **Manufacturing Operation Assignment** | | |
| Process Plan ID | N[7] | Unique identifier for the process plan. |
| Manufacturing Operation ID | N[7] | Unique identifier for the manufacturing operation. |
| Manufacturing Operation Description | C[60] | Brief textual description. |
| Preceeding Manufactuirng Operation ID | N[7] | Unique identifier for preceeding manufacturing operation. |
| Next Manufacturing Operation ID | N[7] | Unique identifier for next manufacturing operation to be performed. |
| Alternative Manufacturing Operation ID | N[7] | Unique identifier for alternative manufacturing operation. |
| Setup time per unit item (min) | N[6,2] | Time required to equip and prepare the 'work centre' or cell for production. |
| Machining time per unit item (min) | N[6,2] | Actual productive time for manufacture of part. |
| Handling time per unit item (min) | N[6,2] | Time required for handling of part which includes transportation. |
| Operation time (min) | N[10,2] | Cumulative setup, machining and handling times for the manufacture of the order. |
| Scrap rate | N[5,2] | The percentage difference between the amount or number of unit parts started in a manufacturing process and that amount or number of units which is completed at an acceptable quality level. |
| **Manufacturing Facility Assignment** | | |
| Manufacturing Operation ID | N[7] | Unique identifier for the manufacturing operation. |
| Asset ID | N[7] | Unique identifier for either an employed personnel or an item which is owned by the business and has value that can be measured objectively. |
| Feed | N[4] | Machining feedrate. |
| Speed (mm/min) | N[6] | Cutting speed. |
| Depth of Cut (mm) | N[3] | |
| Number of Passes | N[5] | |
| Remarks | C[60] | Textual comment. |
| **Resource Assignment** | | |
| Manufacturing Operation ID | N[7] | Unique identifier for the manufacturing operation. |
| Resource ID | C[15] | Unique identifier assigned to resource item |
| Resource Type | C[2] | Classification of resources :<br>T = Toolings<br>TA = Tooling Accessories or Attachments<br>M = Materials<br>F = Fixturing Elements<br>FA = Fixturing Accessories or Attachments<br>M = Miscellaneous |
| Quantity Required | N[12,2] | Total quantity of resource item to be allocated or reserved. |
| Unit of Measure | C[4] | It is the standard quantitive unit for the part used in the manufacturing process. |

WIP

| FIELD DEFINITION | SIZE | DESCRIPTION |
|---|---|---|
| Manufacturing Order Number | N[7] | User defined unique identifier for a batch of manufacturing order. |
| Part Number | C[15] | Unique identifier for the part (or component). |
| Actual Quantity Produced | N[9] | The number of parts manufactured. |
| Work Centre or Cell Utilization Rate | N[5,2] | The percent time that a 'work centre' or cell is running production. |
| Actual capacity (hours) | N[5,2] | Capacity calculated from actual performance data, i.e. number of parts produced multiplied by standard hours per part. |

Order Entry

| FIELD DEFINITION | SIZE | DESCRIPTION |
|---|---|---|
| Customer ID | N[7] | Unique identifier assigned to a customer. |
| Preceeding Manufacturing Order Number | N[7] | Unique identifier assigned for the previous batch of manufacturing order. |
| Current Manufacturing Order Number | N[7] | Unique identifier assigned for the current batch of manufacturing order. |
| Parent Part Number | C[15] | Unique identifier for the parent item which is the higher level item in the BOM. |
| Description | C[60] | Brief textual description of the customer order/product. |
| Product Effectivity Start Date | Date | Date from which the product is being supported. |
| Product Effectivity End Date | Date | Date from which the product is no longer supported. |
| Type (Repeat/One Off) | C[1] | Classification of manufacturing order for either repeat or one off type. |
| Due Date | Date | Planned completion or shipment date for the product. |
| Unit of Measure | C[4] | It is the standard quantitive unit for the part used in the manufacturing process. |
| Unit Price | N[12,2] | It is the price per unit of the item being ordered. |
| Order Quantity | N[9] | Number of items ordered at the specified unit of measure. |

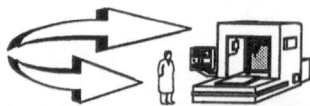

**Manufacturing Facility** — Applies to both machines and manpower requirements

| FIELD DEFINITION | SIZE | DESCRIPTION |
|---|---|---|
| Asset ID | N[7] | Unique identifier for either a personnel or an item which is owned by the business and has value that can be measured objectively. |
| Description | C[60] | Brief textual description for the item. |
| Location | C[15] | Physical location where the personnel is assigned or where the asset item can be found or is being currently used. |
| Working Capacity (hours) | N[3] | Allocated productive time available for working. |
| Labour Cost / hour | N[7,2] | |
| Handling Cost / hour | N[7,2] | |

**Machine & Work parameters**

| FIELD DEFINITION | SIZE | DESCRIPTION |
|---|---|---|
| Asset ID | N[7] | |
| Supplier ID | N[7] | Unique identifier for supplier of asset item. |
| Last Service / Maintenance Date | Date | Date on which service or maintenance is carried out on the asset item. |
| Repaired on | Date | Date on which repair work was carried out on the asset item. |
| Repair Work Order Number | N[7] | Work order number for repair work on the asset item. |
| Maximum Job Size Accommodated | | Physical size of part that can be handled without any problems by the asset item. |
| X axis (mm) | N[5,2] | |
| Y axis (mm) | N[5,2] | |
| Z axis (mm) | N[5,2] | |
| Accuracy | N[4,2] | The degree of freedom from error. |
| **Standard Cost** | | |
| Machining cost / hr | N[7,2] | |
| **Utilities** | | |
| Horse power | N[7] | |
| Speed Range Maximum (mm/min) | N[6] | |
| Minimum (mm/min) | N[6] | |
| Feed Range Maximum (mm/min) | N[4] | |
| Minimum (mm/min) | N[4] | |
| Payload (kg) | N[5] | |
| **Work Envelope** | | |
| X axis (mm) | N[5,2] | |
| Y axis (mm) | N[5,2] | |
| Z axis (mm) | N[5,2] | |
| A axis (mm) | N[5,2] | |
| B axis (mm) | N[5,2] | |
| **Standard Times** | | |
| Setup Time (min) | N[5,2] | |
| Tool Change Time (min) | N[5,2] | |
| Feed Change Time (min) | N[5,2] | |
| Speed Change Time (min) | N[5,2] | |
| Table Rotation Time (min) | N[5,2] | |
| Tool Adjustment Time (min) | N[5,2] | |
| Rapid Traverse (mm/min) | N[5,2] | |

**Personnel**

| FIELD DEFINITION | SIZE | DESCRIPTION |
|---|---|---|
| Asset ID | N[7] | |
| Personnel ID | C[15] | |
| Name | C[30] | |
| Salary | N[7,2] | |
| Address | C[60] | |
| Telephone | C[20] | |
| Skill | C[30] | Skills or expertise personnel possess. |
| Skill Level | N[2] | |
| Remarks | C[60] | |

Resource — Applies to Tools/Materials/Fixtures

| FIELD DEFINITION | SIZE | DESCRIPTION |
|---|---|---|
| Resource ID | C[15] | Unique identifier assigned to resource item. |
| Resource Type | C[2] | Classification of resources:<br>T = Toolings<br>TA = Tooling Accessories or Attachments<br>M = Materials<br>F = Fixturing Elements<br>FA = Fixturing Accessories or Attachments<br>M = Miscellaneous |
| Description | C[60] | Brief textual description for resource item. |
| Location | C[15] | Storage or usage area where the resource item can be found. |
| Account Number | N[15] | Assigned number for purchase of the item. |
| Unit of Measure | C[4] | It is the standard quantitive unit for the part used in the manufacturing process. |
| Unit Price | N[12,2] | It is the price per unit of the item being ordered. |
| Buy/Make/Supply Code | C[1] | It indicates if the item is as follows :<br>M = Make (Manufactured in-house)<br>B = Buy (Purchased and no parts need to be supplied to the vendor)<br>S = Supplied (Purchased but supplied to the vendor) |
| Supplier ID | N[7] | Unique identifier assigned to the supplier of the resource item. |
| Catalogue Order Number | C[30] | Catalog number for the supplied resource item. |
| Purchasing Lead Time | N[7] | The span of time required to obtain a purchased item which includes procurement lead time, vendor lead time, transportation time, receiving, inspection and put away time. |
| Last Order Date | Date | Date on which the last order was placed for the resource item. |
| Quantity Ordered | N[9] | Number of items ordered. |
| Effectivity Start Date | Date | Date from which the resource item is being supported. |
| Effectivity End Date | Date | Date from which the resource item is no longer supported. |
| Stock On-hand | N[9] | Physical stock on-hand minus allocations, reservations and (usually) quantities held for quality problems. |
| Allocated/Reserved Stock | N[9] | Committed resource item. |
| Scrap Value | N[7,2] | Value of scrap per unit measure of scrap. |
| Unit of Measure for Scrap | C[3] | It is the standard quantitive unit for the scrap item. |

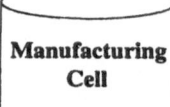  Grouping of manufacturing stations into cells for part manufacture.

| FIELD DEFINITION | SIZE | DESCRIPTION |
|---|---|---|
| Manufacturing Cell Group ID | N[2] | Unique identifier for manufacturing cell. |
| Number of Manufacturing Stations | N[2] | Number of manufacturing stations or processes supported in the configuration. |
| Manufacturing Station 1 - Assest ID | N[7] | Unique identifier for manufacturing station. |
| Description | C[60] | Brief textual description for manufacturing activity. |
| Manufacturing Station 2 - Asset ID | N[7] | " |
| Description | C[60] | . |
| Manufacturing Station 3 - Asset ID | N[7] | " |
| Description | C[60] | |
| Manufacturing Station 4 - Asset ID | N[7] | " |
| Description | C[60] | |
| Manufacturing Station 5 - Asset ID | N[7] | " |
| Description | C[60] | |

| FIELD DEFINITION | SIZE | DESCRIPTION |
|---|---|---|
| Customer ID | N[7] | Unique identifier assigned to a customer. |
| Company/Name | C[40] | Name of customer. |
| Address | C[60] | |
| Contact Person | C[25] | |
| Telephone | C[20] | |
| Fax | C[20] | |

| FIELD DEFINITION | SIZE | DESCRIPTION |
|---|---|---|
| Supplier ID | N[7] | Unique identifier assigned to a supplier. |
| Company/Name | C[40] | Name of supplier. |
| Address | C[60] | |
| Contact Person | C[25] | |
| Telephone | C[20] | |
| Fax | C[20] | |

# Appendix II

*Proprietary information representation in MCC and ELMS*
*CAPM software packages and their correlation to the information models*

**Schedule data tables in MCC**

# Appendix III

## OVERVIEW OF IDEF$_0$ AND IDEF$_{1X}$

### IDEF$_0$

IDEF$_0$ is the technique for modelling functions or activities of the enterprise. It is a descendant of the Structured Analysis and Design Technique (SADT) developed by Ross (1977).

As illustrated, the building block of this modelling approach is the activity box. The activity box defines an activity, or function in the enterprise that is being modelled. The activity may be a decision-making, or information conversion activity, or it may be a material conversion activity, or both. Inputs to the activity are shown at the left of the box. Inputs are items (material, informational) that are transformed by the activity. Outputs of the activity are shown at the right of the box. Outputs are the results of the activity acting on the inputs. Controls are shown entering the activity box from the top. A control is a condition that governs the performance of the activity. For example, a control may be a set of rules governing the activity or a condition that must exist before the activity can be done. Mechanisms enter the activity box from below. A mechanism is the means by which an activity is realised. For example, a mechanism may be a machine, a worker or any enabling element (Figure A).

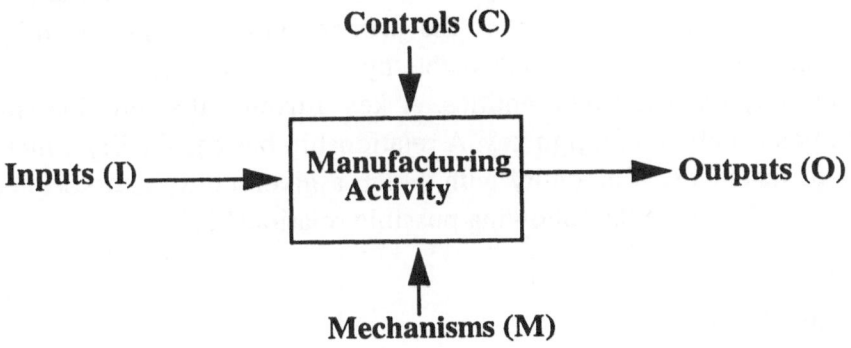

**Figure A** The activity box ICOMs

IDEF$_0$ is applied using top down hierarchic decomposition. At the top of the hierarchy is the overall purpose of the model; it is the global activity that is the subject of the model. The overall activity is decomposable into components that, when taken together, comprise the global activity. This is the second tier of the architecture. Similarly, the second tier activities may be further decomposed into component activities. The decomposition process continues until there is sufficient detail to serve the purpose of the model builder (Figure B).

# IDEF$_{1X}$

IDEF$_{1X}$ is an extension of IDEF$_1$ and is for diagramming the information architechure. It is a semantic data modelling technique that defines the meaning of data within the context of its interrelationship with other data. IDEF$_{1X}$ uses the entity relationship approach based on the ER technique developed by Chen (1977). A completed IDEF$_{1X}$ diagram is a static structure that defines information groupings and relationships among groupings.

As illustrated in Figure C, the basic diagrammatic structure comprises boxes, which are used to represent entities. An entity is a set of real or abstract things (people, object, events) which have common attribute or characteristic. An attribute of an entity set, for which each instance must have a unique value is called a key attribute for that entity.

In the IDEF1X diagram, entity attributes are listed with the box representing the entity. The key attribute is known as the primary key of a given entity and are separated from the rest of the attributes by a line that goes across the box. Relationships may exist between entities. A key attribute that provides the linkage between entities is called a foreign key. A relationship has cardinality which specifies the number of instances of an entity with which a given entity is associated through the relationship. There are the following possible relationships:

- one-to-one
- one-to-many
- many-to-many

Each of the entities becomes a table in the database implementation. The set of attributes of each entity becomes an attribute field (or record field) of the entity table.

**Figure B** Hierarchical decomposition of IDEF$_0$ model

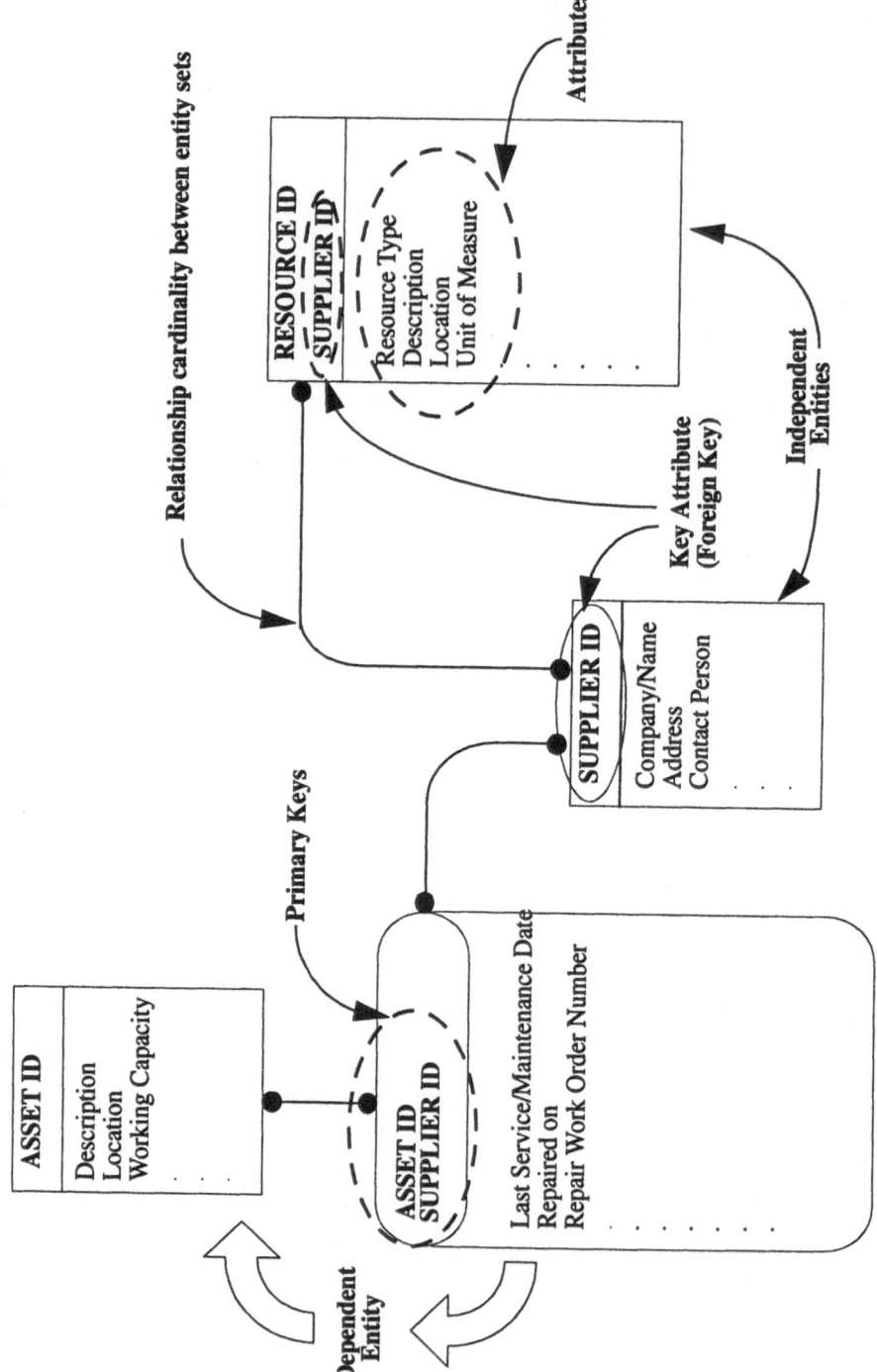

**Figure C** IDEF$_{1X}$ Entity-Attribute relationship

# Appendix IV

## EXPRESS BASED INFORMATION MODEL SCHEMA

### SCHEMA INFORMATION_MODELS;

```
TYPE
type_class = enumeration of (repeat_order, one_off);
END_TYPE;
TYPE
part_class = enumeration of (normal, phantom, resource, co-product, tool, tool_return_item);
END_TYPE;
TYPE
resource_class = enumeration of (tool, tool_assessories, materials, fixtures,
fixture_assessories, miscellaneous);
END_TYPE;
TYPE
order_category = enumeration of (quote_forecast, open_order, confirmed_order,
closed_order_proceed,order_complete, closed_order_archive, order_purge, order_hold);
END_TYPE;
TYPE
 product_code = enumeration of (make, buy, supply);
END_TYPE;


ENTITY ORDER_ENTRY;
has_PART_MASTER-BOM : PART_MASTER-BOM;
has_CUSTOMER : CUSTOMER;
has_SCHEDULE : LIST [1:?] OF SCHEDULE;
has_SHOP_FLOOR_STATUS : LIST [1:?] OF SHOP_FLOOR_STATUS;
Mfg_Order_Number : INTEGER(7);
Preceeding_Order_Number : INTEGER(7);
Description : STRING(60);
Effectivity_Start_Date : date;
Effectivity_End_Date : date;
Type : type_class;
Due_Date : date;
Unit_of_Measure : STRING(4);
Unit_Price : REAL;
Order_Quantity : INTEGER(9);
END_ENTITY;
```

ENTITY PART_MASTER-BOM;
has_BOM_CHILD : LIST [1:?] OF BOM_CHILD;
Parent_part_number : STRING(15);
Effectivity_Start_Date : date;
Effectivity_End_Date : date;
Unit_of_Measure : STRING(4);
Engineering_Change_Notice : INTEGER(7);
Change_Effected_by : STRING(20);
Date_of_Change : date;
Phases_out_Part_Number : STRING(15);
Phased_out_by_Part_Number : STRING(15);
Number_of_Levels : INTEGER(2);
Number_of_components : INTEGER(3);
END_ENTITY;

ENTITY PROCESS_PLAN;
has_BOM_CHILD : BOM_CHILD;
has_MFG_OPERATION_ASSIGNMENT : MFG_OPERATION_ASSIGNMENT;
has_MFG_CELL_CONFIGURATION : MFG_CELL_CONFIGURATION;
Process_plan_ID : INTEGER(7);
Process_Description : STRING(60);
END_ENTITY;

ENTITY BOM_CHILD;
has_ENGINEERING_RESOURCE : LIST [1:?] OF ENGINEERING_RESOURCE;
Part_number : STRING(15);
Number_of_components : INTEGER(3);
Part_Type : part_class;
Quantity_per_Assembly : REAL;
Effectivity_Start_Date : date;
Effectivity_End_Date : date;
Unit_of_Measure : STRING(4);
Lead_Time_Offset : INTEGER(9);
Engineering_Change_Notice : INTEGER(7);
Change_Effected_by : STRING(20);
Change_Date : date;
Phases_Out_Part_Number : STRING(15);
Phased_Out_by_Part_Number : STRING(15);
END_ENTITY;

```
ENTITY MFG_OPERATION_ASSIGNMENT;
has_RESOURCE_ASSIGNMENT : LIST [1:?] OF RESOURCE_ASSIGNMENT;
has_MFG_FACILITY_ASSIGNMENT : LIST [1:?] OF MFG_FACILITY_ASSIGNMENT;
Mfg_Operation_ID : INTEGER(7);
Mfg_Operation_Description : STRING(60);
Preceed_Mfg_Operation_ID : INTEGER(7);
Next_Mfg_Operation_ID : INTEGER(7);
Alternate_Mfg_Operation_ID : INTEGER(7);
Setup_time_per_item : REAL;
Machining_time_per_item : REAL;
Handling_time_per_item : REAL;
Operation_time_per_item : REAL;
Scrap_rate : REAL;
END_ENTITY;

ENTITY RESOURCE_ASSIGNMENT;
has_RESOURCE : RESOURCE;
Resource_Type : resource_class;
Quantity_Required : REAL;
Unit_of_Measure : STRING(4);
END_ENTITY;

ENTITY MFG_FACILITY_ASSIGNMENT;
has_MANUFACTURING_FACILITY : MANUFACTURING_FACILITY;
Feed : INTEGER(4);
Speed : INTEGER(6);
Depth_of_cut : INTEGER(3);
Number_of_passes : INTEGER(5);
Remarks : STRING(60);
END_ENTITY;

ENTITY MFG_CELL_CONFIGURATION;
has_MANUFACTURING_FACILITY : LIST [1:?] OF MANUFACTURING_FACILITY;
Mfg_Cell_Group_ID : INTEGER(2);
Number_of_Mfg_Stations : INTEGER(2);
Mfg_station_1 : INTEGER(7);
Description_station_1 : STRING(60);
Mfg_station_2 : INTEGER(7);
Description_station_2 : STRING(60);
Mfg_station_3 : INTEGER(7);
Description_station_3 : STRING(60);
Mfg_station_4 : INTEGER(7);
Description_station_4 : STRING(60);
Mfg_station_5 : INTEGER(7);
Description_station_5 : STRING(60);
END_ENTITY;
```

```
ENTITY RESOURCE;
Resource_ID : STRING(15);
Resource_Type : resource_class;
Description : STRING(60);
Location : STRING(15);
Account_Number : INTEGER(15);
Unit_of_Measure : STRING(4);
Unit_Price : REAL;
Buy_Make_Supply_Code : product_code;
Catalogue_Order_Number : STRING(30);
Purchasing_Lead_Time : INTEGER(7);
Last_Order_Date : date;
Quantity_Ordered : INTEGER(9);
Effectivity_Start_Date : date;
Effectivity_End_Date : date;
Stock_on_hand : INTEGER(9);
Allocated_Reserved_Stock : INTEGER(9);
Scrap_Value : REAL;
Scrap_unit_of_Measure : STRING(4);
END_ENTITY;

ENTITY ENGINEERING_RESOURCE;
has_PROCESS_PLAN : PROCESS_PLAN;
Engineering_Resource : STRING(10);
Location : STRING(10);
END_ENTITY;

ENTITY CUSTOMER;
Customer_ID : INTEGER(7);
Company_Name : STRING(40);
Address : STRING(60);
Contact_Person : STRING(25);
Telephone : STRING(20);
Fax : STRING(20);
END_ENTITY;

ENTITY SCHEDULE;
has_SHOP_FLOOR_STATUS : SHOP_FLOOR_STATUS;
has_BOM_CHILD : BOM_CHILD;
Priority : INTEGER(3);
Order_Status : order_category;
Planned_Quantity : INTEGER(9);
Unit_of_Measure : STRING(4);
Schedule_Start_Date : date;
Schedule_End_Date : date;
END_ENTITY;
```

```
ENTITY SHOP_FLOOR_STATUS;
has_BOM_CHILD : BOM_CHILD;
Actual_Quantity_Produced : INTEGER(9);
Station_Utilisation_Rate : REAL;
Actual_Capacity_Utilised : REAL;
END_ENTITY;

ENTITY MANUFACTURING_FACILITY
SUPERTYPE OF (ONEOF(PERSONNEL, MACHINE));
Asset_ID : INTEGER(7);
Description : STRING(60);
Location : STRING(15);
Working_Capacity : INTEGER(3);
Labor_Cost_per_hour : REAL;
Handling_Cost_per_hour : REAL;
END_ENTITY;

ENTITY SUPPLIER;
has_RESOURCE : LIST [1:?] OF RESOURCE;
Supplier_ID : INTEGER(7);
Company_Name : STRING(40);
Address : STRING(60);
Contact_Person : STRING(25);
Telephone : STRING(20);
Fax : STRING(20);
END_ENTITY;

ENTITY PERSONNEL
SUBTYPE OF (MANUFACTURING_FACILITY);
Personnel_ID : STRING(15);
Name : STRING(30);
Address : STRING(60);
Telephone : STRING(20);
Salary : REAL;
Skill : STRING(30);
Skill_level : INTEGER(2);
Remarks : STRING(60);
END_ENTITY;
```

```
ENTITY MACHINE
SUBTYPE OF (MANUFACTURING_FACILITY);
has_SUPPLIER : SUPPLIER;
Last_Service_Maintenance_Date : date;
Repaired_on : date;
Repair_Work_Order_Number : INTEGER(7);
Max_job_size_X_axis   : REAL;
Max_job_size_Y_axis  : REAL;
Max_job_size_Z_axis  : REAL;
Accuracy : REAL;
Machining_Cost_per_hour : REAL;
Horse_Power : INTEGER(7);
Speed_Range_Min : INTEGER(6);
Speed_Range_Max : INTEGER(6);
Feed_Range_Min : INTEGER(4);
Feed_Range_Max : INTEGER(4);
Payload : INTEGER(5);
Working_Envelope_X_axis  : REAL;
Working_Envelope_Y_axis  : REAL;
Working_Envelope_Z_axis  : REAL;
Working_Envelope_A_axis  : REAL;
Working_Envelope_B_axis  : REAL;
Setup_Time : REAL;
Tool_Change_Time : REAL;
Feed_Change_Time : REAL;
Table_Rotation_Time : REAL;
Tool_Adjustment_Time : REAL;
Rapid_Tranverse_Rate : REAL;
END_ENTITY;

ENTITY date;
day  : INTEGER(2);
month : INTEGER(2);
year : INTEGER(2);
END_ENTITY;


END_SCHEMA;
```

# GLOSSARY

**Assemble-to-order**

A product where all components (bulk, semi-finished, intermediate, subassembly, fabricated, purchased, packaging, etc.) used in the assembly, packaging or finishing process are planned and stocked in anticipation of customer order.

**Bespoke system**

A system designed and implemented specifically for a particular (one-off) application or site.

**CAD/CAM**

Computer-aided design/ Computer-aided manufacturing

**CAPM**

Computer-aided production management

**CAPP**

Computer-aided process planning

**CASE**

Computer-aided software engineering (tools) used to speed up and formulate the process of software design. Such systems use a variety of representations such as data flow diagrams, entity-relationship diagrams, and in some cases generate program code.

**CIM-BIOSYS IIS**

Loughborough University of Technology Manufacturing Systems Integration Research Institute's systems integrating infrastructure.

**CIM model**

A representation of a CIM system.

## CIM-OSA

Open systems architecture for CIM — the output of a major ESPRIT project (No. 5288: AMICE) which attempts to formalise the design, implementation and running of 'open' CIM systems.

## Configure-to-order / Engineer-to-order

Products whose customer specifications require unique engineering design or significant customisation. Each customer order then results in a unique set of part numbers, bills of material and routings.

## Database

A mechanised, formally defined, centrally controlled collection of data.

## Database management system (DBMS)

A software system which performs the functions of defining, creating, revising and controlling the database.

## Database prototyping

Technique used to discover the information requirements and to construct a data model.

## Data integrity

The ability of the database to remain correct during operation.

## Data Manipulation Language (DML)

Highly non-procedural languages associated with database system, e.g. 4GL.

## Data store

The set of data storage facilities containing the shared data.

## DCE

Distributed computing environment

## Distributed database

Database spread across a network of computers.

## DNC

Distributed numerical control.

**Driver**

A driver is a customised software interface used to enable the CIM-BIOSYS IIS to incorporate proprietary devices and applications which have their own specific protocols.

**ESPRIT**

The European Strategic Programme for Research and Development in Information Technology

**Essential model**

A primary model or representation (of some aspect of manufacturing systems, objects or process) at a generic level.

**Functional interaction module**

Functional interaction module has been conceived with the purpose of tying together a set of applications into a coherent system to enable software interoperability. It controls and co-ordinates the interaction between applications and also synchronises the flow of activities within the system to support part manufacture.

**IDEF**

ICAM Definition methodology — a set of systems analysis and design tools.

**Information of common interest (shared information)**

Information which is created by one function and is often processed and/or used by several other functions within the enterprise.

**Integration**

The aggregation of resources and applications into a synergistic whole.

**Integration toolset**

A set of complimentary software programs to assist in or enable some aspect of the development or management of CIM systems.

**IRDS**

Integrated resource dictionary system

**Legacy**

The term legacy (systems, components and software) is used to refer to a previously installed base of systems, components and software. Legacy elements will not normally conform to the methods and standards which will be adopted in current generation solutions.

**Life-cycle**

The specification, design, implementation and useful life of a system.

**Make-to-order**

A product which is manufactured after receipt of a customer order. Frequently long lead-time components are planned prior to the order arriving in order to reduce the delivery time to the customer. Where options or other subassemblies are stocked prior to customer orders arriving, the term 'assemble-to-order' is frequently used.

**Make-to-stock**

A product which is shipped from finished goods, 'off the shelf', and therefore is finished prior to a customer order arriving.

**Model**

A model can be defined as a tentative description of a system that accounts for properties relevant to the intended purposes of the model.

**Normalise**

The decomposition of more complex data structures into flat files (relations). This forms the basis of relational databases.

**Open solution**

A solution (to a CIM requirement) which does not constrain the user to specific proprietary hardware, software or protocols.

**Platform (of integration services)**

A software system which provides a consistent set of integration services (interaction, information and configuration) to manufacturing applications, to enable them to perform as part of a CIM system.

**Primary key (or tuple)**
    A key which uniquely identifies a record (or data grouping).

**Product introduction**
    The activities associated with specification, design, analysis, process and resource planning and other functions required to bring a product to the production stage.

**Proprietary**
    Belonging or under the control of a private organisation (e.g. AUTOCAD is a proprietary package, SNA is a proprietary protocol).

**RDA**
    Remote data access

**Relation**
    A two-dimensional array of data elements (implemented as a table in a relational database).

**Relational database**
    A database made up of relations. Its database management system has the capability to form different relations thus giving great flexibility in the usage of data.

**Schema**
    A structured description of the information available in a database.

**SME**
    Small and medium sized enterprises

**Table**
    A collection of data (in a relational database) suitable for quick reference, each item being uniquely identified either by a label or its relative position.

**Turnkey**
    A turnkey system is one which is delivered and implemented by the supplier with little effort on the part of the user. It is largely 'pre-designed' (at least at the component level) and the user sacrifices a close match to his own requirements in order to be able to be 'up and running' quickly.

# REFERENCES

Afferson, M., Andrews, J. K., Muhlemann, A. P., Price, D. H. R., Sharp, J. A., 1992, *Generic manufacturing information systems development via template prototyping*, European Journal of Information Systems, Vol. 1, pp379-386.

Aguiar, M. W., Weston, R. H., 1993, *CIM-OSA and stochastic time Petri nets for behavioural modelling and model handling in CIM systems design and building*, Procs. of the Institution of Mechanical Engineers, Vol. 207. Part B. Journal of Engineering Manufacture, pp147-158.

Aguiar, M. W., Weston, R. H., August 1993, *Reference Architectures for Enterprise Integration*, Procs. of CARS/FOF' 93 Conference, USA.

Akif, H. C., Documeings, G., 1991, *Computer aided GRAI method (C. A. GRAI)*, Procs. of Advances in Production Management Systems - IFIP/91, France, pp283-292.

AMR (Advanced Manufacturing Research), March 1991, *Application Enabler*, Report, USA.

Anscombe, J., November 1992, *Integration - Breaking the Barriers to Excellence*, Procs. Twenty-seventh Annual BPICS Conference, Birmingham, UK, pp89-103.

Arngrimsson, G., Vesterager, J., August 1992, *STEP: Experiences from actual use of the standard*, Procs. of IFIP Working Group 5.7, Conference on Integration in Production Management Systems, Eindhoven, Netherlands, pp23-35

Bailin, S. C., 1989, *An object-oriented requirements specification method*, Communications of the ACM, Vol. 32, No. 5, pp608-623.

Barkmeyer, E. J., 1989, *Some Interactions of information and control in Integrated Automation systems*, Advanced Information Technology, Industrial Material Flow Systems, Springer-Verlag.

Batini, C., Lenzerini, M., Navathe, S. B., December 1986, *A Comparative Analysis of Methodologies for Database Schema Integration*, ACM Computing Surveys, Vol. 18, No. 4, pp322-364.

Bauer, A., 1991, *Shop floor control systems: from design to implementation*, Chapman and Hall.

Beech, D., Ozbutun, C., 1990, *Object Database as generalizations of relational databases*, Proc. of the Object-Oriented Database Task Force Group Workshop, Ottawa, Canada, pp119-135.

Beerit, C., October 1993, *New Directions in Database Management Systems*, Procs. of Fifth Jerusalem Conference on Information Technology, Israel, pp500-506.

Bohse, M. E., Harhalakis, G., 1987, *Integrating CAD and MRP II Systems*, CIM Review, Vol. 3, No. 4, pp7-15.

Bond, T. C., 1993, *An investigation into the use of OPT production scheduling*, International Journal Production Planning and Control, Vol. 4, No. 4, pp399-406.

Breitbart, A., Morales, H., Silberschatz, A., Thompson, G., October 1993, *Multidatabase Concurrency Problems - Multidatabase Transctions Concurrency Control Mechanisms*, Procs. of Fifth Jerusalem Conference on Information Technology, Israel, pp507-519.

Bright, M. W., Hurson, A. R., Pakzad, S. H., 1992, *A Taxonomy and Current Issues in Multidatabase Systems*, IEEE, pp50-59.

Buchman, A. P., 1984, *Current Trends in CAD Databases*, Computer-Aided Design, Vol. 16, No. 3.

Chang, Tien-Chien, 1985, *An Introduction to Automated Process Planning Systems*, Prentice-Hall.

Chaudhri, A. B., 1993, *Object database management systems: an overview,* BCS OOPS Newsletter, No. 8, pp6-15.

Chaudhri, A. B., Revell, N., 1994, *Object database benchmarks: past, present and future*, Proc. of Object-Oriented Databases: Realising their Potential and Interoperability with RDBMS, London, UK.

CIM-OSA ESPRIT Consortium AMICE, 1989, *Open System Architecture for CIM*, Springer-Verlag, Berlin (D).

CIM Strategies, March 1990, *DELTA factory floor manager combines data management methods*, pp7-10.

CIM Strategies, March 1991, *Application Case Study, Interoperability standards form a base for CIM*, Vol. 8, No. 3, pp4-7.

Clements, P., February 1991a, Internal Report on the EXPRESS to SQL Compiler, Loughborough University of Technology, UK.

Clements, P., March 1991b, Internal Report on the STEP Parser, Loughborough University of Technology, UK.

Clements, P., October 1992, *The application of EXPRESS modelling and tools within an integration platform*, Second EXPRESS Users Group, Dallas, USA.

Clements, P., Coutts, I. A., Weston, R. H., September 1993, *A life-cycle support environment comprising open systems manufacturing modelling methods and the CIM-BIOSYS infrastructural tools*, Proc. of the Symposium on Manufacturing Applicaton Programming Language Environment (MAPLE) Conference, Ottawa, Canada, pp181-195.

Clements, P., Hodgson, A., Leech, M., Ryan, A., November 1991, *Information Systems Modelling and Implementation in an industrial environment*, Procs. of AUTOFACT '91, Chicago, IL, USA.

Codd, E. F., 1992, *Dr. Codd on "End of Relational"*, DBMS, Vol. 5, No 11:6.

Colquhoun, G. J.,Baines, R. W., Crossley, R., 1996, *A composite behavioural modelling approach for manufacturing enterprises*, International Journal Computer Integrated Manufacturing, Vol. 9, No. 6, pp463-475.

Colquhoun, G. J., Baines, R. W., Crossley, R., 1993, *A state of the art review of IDEF$_0$,* International Journal Computer Integrated Manufacturing, Vol. 6, No. 4, pp252-264

Cutts, G., 1991, *SSADM Structured Systems Analysis and Design Methodology*, Blackwell Scientific, Oxford, UK.

Czernik, S., Quint, W., 1992, *Selection of methods, techniques and tools for system analysis and for the integration of CIM elements in existing manufacturing organizations*, International Journal Production Planning and Control, Vol. 3, Part 2, pp202-209.

DatabaseBuyer's Guide Supplement, February 1990, *Production Management Software*, Industrial Computing, pp46-58.

DATAPRO, March 1992, *Manufacturing Automation Series: Factory Automation Systems*, McGraw Hill, USA.

Date, J., 1986, *An Introduction to Database systems*, Vol. 1, Addison-Wesley Publishing Co. Inc.

Davis, G. B., Olson, M. H., 1987, *Management Information Systems*, Second Edition, McGraw-Hill, pp502-504.

De Toni, A., Caputo, C., Vinelli, A., 1988, *Production management techniques*, International Journal of Operations and Production Management, Vol. 8, No. 2, pp35-51.

Dettmer, R., November 1995, *A class act - the rise of object-oriented technology*, IEE Review, pp253-256.

De Vaan, M. J., July-September 1992, *Introduction MRP II, with enhancements: the case of a furniture manufacturer*, International Journal Production Planning and Control, Vol. 3, No. 3, pp258-263.

Dinitz, M., July 1990, *Configure-To-Order: An industry challenge*, Industrial Engineering, pp21-22.

Drucker, P. F., November 1991, *The Factory of the Future*, World Executive's Digest, pp26-32.

DTI, 1987, Moore & Matthes Ltd, *UK, Through MAP to CIM*, Department of Trade and Industry, UK.

DTI, 1989, PA Consulting Group, *Manufacturing into the late 1990s*, HMSO, Department of Trade and Industry, UK.

DTI, 1993, *Computer Integrated Manufacturing - A Survey of Worldwide R & D*, Department of Trade and Industry, UK.

ELMS Technical Manual, 1990.

ESPRIT Consortium, 1989, *Open System Architecture for CIM*, Project 688, Vol. 1, Springer-Verlag, pp13-16.

Evans, C. D., Meek, B. L., Walker, R. S., 1993, *User Needs in Information Technology Standards*, Butterworth-Heinemann Ltd (Publisher), UK.

Foong, N. F., Ang, K. P., Singh, V., May 1992, *Computer Simulation as a Tool for Integrated Manufacturing*, Procs. Asia-Pacific Industrial Automation (IA)' 92 Conference, Singapore.

Fritsch, C. A., 1989, *Information Dynamics for Computer Integrated Product Realisation*, NATO ASI Series, Springer Verlag, Vol. F53:, pp21-38.

Fry, T., Karwan, K., Baker, W., 1993, *Performance measurement systems and time-based manufacturing*, International Journal of Production Planning and Control, Vol. 4, No. 2, pp102-111.

Golberg, C. J., Winter 1993, *Object Oriented Databases - The New Wave in RDBMS Technology*, ORACLE, Vol. VII, No. 1, pp35-39.

Goldratt, M., E., 1988, *Computerized shop floor scheduling*, International Journal of Production Research, Vol. 26, No. 3, pp443-455.

Gould, L., August 1992, *CIM Interface Modules: A route to Open Systems*, Managing Automation, Vol. 7, No. 8, pp47-50.

Goyal, S. K., Gunasekaran, T. Martikainen, Yli-Olli, P., 1993, *Design of optimal configuration for a multi-stage production system*, International Journal of Production Planning and Control, Vol. 4, No. 3, pp239-252.

Halevi, G., Weil, R., 1992, *CAPP as concurrent link between Design and Production Management*, IFIP Transactions Part B Applications in Technology, Vol. 6, pp177-184.

Halladay, S., Wiebel, M., 1993, *Object-Oriented Software Engineering*, Lawrence, Kan.: R & D Publications.

Halsall, D. N., Muhlemann, A. P., Price, D. H. R., September 1993, *A Production Planning and Resource Scheduling Model from Small Manufacturing Enterprises*, Procs. Ninth National Conference on Manufacturing Research, UK.

Harhalakis, G., Lin, C. P, Hillion, H. , Moy, K. Y., 1990, *Development of a factory Level CIM Model*, Journal of Manufacturing Systems, Vol. 9, No. 2, pp116-128.

Hars, A., 1990, *CIDAM - modules for the creation of CIM*, Procs. Sixth CIM-Europe Annual Conference, pp286-295.

Hars, A., Heib, R., Kruse, Chr., Michely, J., Scheer, A., -W., May 1992, *Reference Models for Data Engineering in CIM*, Procs. Eighth CIM-Europe Annual Conference, Birmingham, UK, pp249-260.

Hayes, F., Spring 1992, *Esperanto for Databases*, Unixworld-Supplement: Special Report Interoperability, pp49-51.

Higgins, P., Tierney, K., Browne, J., September 1991, *Production Management State of the Art and Perspectives*, Procs. Fourth International IFIP TC5 Conference, Computer Applications in Production and Engineering, Bordeaux, France, pp3-14.

Himes, D. A., 1993, *Database interoperability and portability through standards*, Procs. of the Second International Conference on Parallel and Distributed Information Systems, pp225-256.

Hind, C. J., West, A. A., Williams, D., J., 1990, *The use of object orientation for the design and implementation of manufacturing process control systems*, Internal Report, Dept of Manufacturing Engineering, Loughborough University of Technology, LUT Press, UK.

Hodgson, A., 1993, *Production Planning and Control within a CIM environment: some current developments and requirements for the future*, International Journal Production Planning and Control, Vol. 4, No. 4, pp296-303.

Hodgson, A., Waterlow, G., 1992, *Special feature: Computer-aided production management*, Computing & Control Engineering Journal, IEE, ISSN 0956-3385, Vol. 3, No. 2.

Hodgson, A., Weston, R. H., 1993, *Application and Information Support Systems for Planning and Control in CIM*, Grant No. GR/F 69192, ACME Review Final Report, UK.

Hodgson, A., Weston, R. W., Sumpter, C. M., Gascoigne, A., August-September 1988, *Planning And Control Information flow in CIM*, Procs. International Conference on Factory 2000 - Integrating Information and Material Flow, Cambridge, UK, pp49-56.

Hollyman, B., Anderson, L., January 1991, *Implementing an Open Systems Architecture*, CommUNIXations, Published by Uniforum (International Association of Unix Systems Users), Vol. XI, No. 1, pp23-29.

Hugh Ujhazy, March/April 1995, *Relational Object Bases*, ORACLE Magazine, Vol. IX, No. 2, pp81-83.

Hughes, D., August-September 1988, *Criteria for the distribution of information processing in factory 2000*, Procs. International Conference on Factory 2000 - Integrating Information and material flow, Cambridge, England, pp45-48.

ICAM, December 1985, *Information Modelling Manual IDEF1 - Extended*, ICAM Project Report (Priority 6201), D. Appleton Company, Inc, Manhattan Beach, CA, USA.

ISO, 1991, *MANDATE*, ISO TC184/SC4/WG8 Document N1, ISO TC184/SC4 Secretariat, National Institute of Standards and Technology, Gaithersburg, MD 20899, USA.

ISO, 1994, ISO DIS 10303-1, *Industrial automation systems and integration -Product Data Representation and exchange Part 1: Overview and Fundamental Principles*, International Organization for Standardization, Geneva.

ITAP Technology Seminar, 1990, *Advances in Computer Integrated Manufacturing*, ITAP Technology Report No. 5/90, National Computer Board, Singapore.

Jain, K. H., Bu-Hulaiga, I. M., Summer 1992, *E-R Approach to Distributed Heterogeneous Database Systems for Integrated Manufacturing*, Journal of Database Administration, Vol. 3, Part 3, pp21-29.

Jeng, B. C., Chao, W. S., July 1992, *Communicating Objects for System Integration modelling*, Procs. Second International Conference on Automation Technology, Taipei, Taiwan, Vol. 2, pp307-312.

Jochem, R., 1989, *An object oriented analysis and design methodology for computer integrated manufacturing systems*, Tools 89, pp75-84.

Jones, G., Roberts, M., 1990, *Optimized Production Technology (OPT)*, IFS Publications, UK.

Joris, S. M., Vergeest, Matthijis, Sepers, June 1993, *Techniques to make CAD/CAM Systems communicative*, Procs. of the Third International Flexible Automation and Integrated Manufacturing, University of Limerick, Ireland, pp255-266.

Jorysz, H. R., Vernadat, F. B., 1990, *CIM-OSA part 1: Total enterprise modelling and function view*, International Journal Computer Integrated Manufacturing, Vol. 3, Nos 3 and 4, pp144-156.

Kelli Wiseth, January/February 1995, *Data Warehousing - Architecture for the Information Age*, ORACLE Magazine, Vol. IX, No. 1, pp34-37.

Khoshafian, S., Blumer, R., Abnous, R., 1990, *Inheritance and generalization in Intelligent SQL*, Proc. of the Object-Oriented Database Task Force Group Workshop, Ottawa, Canada, pp103-118.

Kochhar, A. K., Monniott, J. P., Price, D. H. R, Rhodes, D. J., Towill, D. R., Waterlow, J. G., 1987, *A study of computer aided production management in UK batch manufacturing*, International Journal of Operations and Production Management, Vol. 7, pp7-57.

Kong, M., December 1995, *DCE: An Environment for Secure Client/Server Computing*, Hewlett-Packard Journal, Vol. 46, No. 6, pp6-15.

Koriba, M., 1983, *Database Systems: Their Applications to CAD Software Design*, Computer-Aided Design, Vol. 15, No. 5.

Kosanke, K., 1991, *Open Systems Architecture for CIM (CIM-OSA) Standards for Manufacturing*, Procs. International Conference on Computer Integrated Manufacturing (ICCIM' 91), Singapore.

Krishnamurthy, R., Litwin, W., Kent, W., April 1991, *Interoperability of Heterogeneous Databases with schematic discrepancies*, Procs. First International Workshop on Interoperability in Multidatabase systems, Kyoto, Japan, pp144-151.

Lang-Lendroff, G., Unterburg, J, June 1989, *Changes in understanding of CAD/CAM: a database-oriented approach*, Computer Aided Design, Vol. 21, No. 5, pp309-314.

Lars, D. T., 1990, *Is there a "GAP" of knowledge between R&D and Production?*, Advances in production management systems, Procs. Fourth International IFIP Conference TC5/WG 5-7, Espoo, Finland.

Larsen, N. E., Alting, L., 1993, *Criteria for selecting a production control philosophy*, International Journal Production Planning and Control, Vol. 4, No. 1, pp54-68.

Lee, C. Y., 1993, *A Recent Development of the Integrated Manufacturing System: A Hybrid of MRP and JIT*, International Journal of Operations and Production Management, Vol. 13, No. 4, pp3-17.

Lim, B. S., July-October 1992, *CIMIDES - A Computer Integrated Manufacturing Information and Data Exchange System*, International Journal of Computer Intergrated Manufacturing, Vol. 5, Nos. 4 & 5, pp240-254.

Logan, F. A., March 1986, *Evolutionary Cycle of an Expert CAPP System*, Procs. Conference CIMTECH, Boston, MA, USA.

Lopes, P. F., 1992, *CIM II: The Integrated Manufacturing Enterprise*, Industrial Engineering, Vol. 24, No. 11, pp43-45.

Luscombe, M., 1991, *Design and Implementation of Integrated Production Control systems*, Integrated Manufacturing System, Vol. 2, No. 4, pp4-8.

Maier, D., 1989, *Object-Oriented Concepts, Databases, and Applications*, Edited by W. Kim and F. H. Lochovsky, Addision-Wesley.

Maji, R. K., October 1988, *Tools for development of Information Systems in CIM*, Advanced Manufacturing Engineering, Vol. 1, pp26-34.

Martin, J., 1980, *Computer Data Base Organization*, Second Edition, Prentice-Hall, Englewood Cliffs, NJ, USA.

Maude. T., Willis, G., 1991, *Rapid Prototyping,* Pitman Publishing, London, UK.

Maull, R. S., Childe, S. J., 1993, *A step-by-step guide to the identification of an appropriate computer-aided production management system*, International Journal of Production Planning and Control, Vol. 4, No. 1, pp69-76.

Mayer, R. J., Painter, M. K., 1991, *Roadmap for enterprise integration*, Procs. of Autofact 91, USA.

MCC Technical Manual, 1989, John Brown Systems PLC, UK.

Meta Software, 1990, Design/IDEF User's Manual, Meta Software, MA, USA.

Metz, S., August 1990, *Making Manufacturing Better, not just faster,* Managing Automation.

Moerman, P.A., 1991, *The evaluation of technology in relation to products and markets: observations, considerations, experience, and solutions*, International Journal of Computer Integrated Manufacturing, Vol. 4, No. 1, pp2-15.

Motro, A., July 1987, *Superviews: Virtual Integration of Multiple Databases*, IEEE Trans. Software Engineering, Vol. 13, No. 7, pp785-798.

Mowbray, T. J., Zahavi, R., 1995, *The Essential CORBA*, John Wiley & Sons, Inc.

Muhlemann, A. P., Price, D. H. R., Sharp, J. A., Afferson, M., 1991, *Fourth Generation languages and integrated information systems for small manufacturing companies*, International Journal Computer Integrated Manufacturing, Vol. 4, No. 1, pp16-22.

Muhlemann, A. P., Price, D. H. R., Sharp, J. A., Afferson, M., Andrews, J. K., 1990, *Information systems for use by production managers in smaller manufacturing enterprises*, Procs. of the Institution of Mechanical Engineers (Part B), Vol. 204, p191-196.

Olle, T. W., 1978, *The CODASYL Approach to Database Management Systems*, John Wiley and Sons, New York.

ORACLE, 1996, Database set of manuals, Version 7.0, Oracle Corporation.

Orfali, R., Harkey, D., Edwards, J., 1996, *The Essential Distributed Objects Survival Guide*, John Wiley & Sons, Inc.

Orr, K., Gane, C., Yourdon, E., Chen, P. P., Constantine, L. L., April 1989, *Methodology: The Experts Speak*, Byte, pp221-244.

Paranuk, H. V. D., 1988, *Chapter 5: Factory communication system, Artificial Intelligence: Implications for CIM*, IFS Publications Ltd, Springer-Verlag.

Perkovic, P., Spring 1991, *SQL Access and ANSI/ISO SQL and X/Open*, COMPCON, pp120-122.

Peters, T., 1989, *Thriving on Chaos*, Pan Books Ltd, UK.

Pheasey, D., November 1992, *Competitive Manufacturing - 'A Vision of the year 2001'*, Procs. Twenty-seventh Annual BPICS Conference, Birmingham, UK, pp23-31.

Plenert, G., 1993, *An Overview of JIT*, International Journal of Advanced Manufacturing Technology, Vol. 8, pp91-95.

Preece, J., 1993, *A Guide to Usability*, Addison-Wesley.

Ptak, C. A., 1991, *MRP, MRP II, OPT, JIT, and CIM - Succession, Evolution, or necessary combination?*, Production and Inventory Management Journal, Vol. 32, Part 2, pp7-11.

Pugh, D. S., Hickson, D., J., 1989, *Writers on Organization*, Penguin Books (Fourth Edition), pp90-93.

Rembold, U., Nnaji B. O., Storr, A., 1993, *CIM*, Addison-Wesley, UK.

Robertson, B., July 1996, *The Network Guide to Middleware*, Windows Sources Australia, pp193-202.
Ross, D. T., 1977, *Structured Analysis (SA): A language for Communicating Ideas,* IEEE Transactions on Software Reliability, Vol. 3, No. 1.

Rui, A., 1989, *Information support systems for the distributed planning and control in batch manufacture*, PhD Thesis, Dept. of Manufacturing, Loughborough University of Technology, UK.

Rumbaugh, J., 1991, *Object-Oriented Modeling and Design*, Prentice Hall International.

Rusinkiewicz, M., Czejdo, B., 1987, *An approach to query processing in federated database systems*, Procs. Hawaii International Conference on Systems Sciences.

Sanders, L., Mayer, R. J., Browne, D. C., Menzel, C., 1991, *Containers objects : a description based knowledge representation scheme*, Procs. of Autofact' 91, USA, pp7.39-7.50.

Savolainen, T., 1991, *CIMVIEW: a tool for symbolic top-down simulation for CIM*, Procs. of Advances in Production Management Systems, IFIP, Holland.

Saxe, K., November 1985, *MRP II Into CIM : The Interface Phase*, Procs. Conference Autofact '85, Detroit, MI, USA.

Scheer, A.-W., 1988, *Computer Integrated Manufacturing - Computer Steered Industry*, First Edition, Springer-Verlag.

Scheer, A.-W., 1989, *Enterprise-Wide Data Modelling - Information Systems in Industry*, Springer-Verlag, p259.

Scheer, A.-W., 1991, *CIM - Towards the Factory of the Future*, Second Edition, Springer-Verlag.

Schenck, D., December 1989, *Information Modelling Language EXPRESS*, ISO TC184/SC4/WG1 N442.

Schiel, U., Mistrik, I., 1990, *Using object-oriented analysis and design for integrated systems*, Procs. of the First International Conference on Systems Integration, USA, pp125-134.

Schnur, J. A., Summer 1987, *Can there be CIM Without MRP II?*, CIM Review, USA.

Schonewolf, W., Langendoen, M., Gransier, T., Baisch, R., Drossopoulos, May 1992, *Application of CIM-OSA in Machine Tool Manufacturing and Aluminium Casting*, Procs. Eighth Annual CIM-Europe Conference, Birmingham, UK, pp217-229.

Shaharoun, A. M., Hodgson, A., Weston, R. H., August 1992, *Cost modelling in Advanced Manufacturing Systems*, Procs. of International Conference for Manufacturing Automation (ICMA), Hong Kong.

Shunk, D., Sullivan, B., Cahill, J., Fall 1986, *Making the Most of IDEF Modeling - The Triple Diagonal Concept*, CIM Review, USA, pp2-17.

SI (Systems Integration) Group (LUT), February 1994, *Model Driven CIM: The design, implementation and management of Open CIM systems*, Loughborough University of Technology, UK, SERC/ACME Review Report No. 2, Grant No. GR/H/22798.

SIM, 1993, *User and Technical Manual*, MSPL Ltd., Windsor, UK.

Singh, V., October 1991, *CIM Model for Metal Machining Trade - Translating Vision into Reality*, Procs. International Conference on Computer Integrated Manufacturing (ICCIM' 91), Singapore, pp336-341.

Singh, V., May 1992, *Flexible Materials Handling and Storage System for Integrated Manufacture*, Procs. Asia-Pacific Industrial Automation (IA)' 92 Conference, Singapore, pp10-21.

Singh, V., Weston, R. H., September 1993, *New Generation of "Open" Manufacturing Control Systems for "Seamless" Integration in CIM*, Procs. International Conference on Computer Integrated Manufacturing (ICCIM' 93), Singapore, pp309-321.

Singh, V., Weston, R. H., 1994a, *Functional interaction management: A requirement for software interoperability*, Procs. of the Institution of Mechanical Engineers, Part B, Journal of Engineering Manufacture, UK, pp289-305.

Singh, V., Weston, R.H., May 1994b, *Software Interoperability for Integrated Manufacturing, A Reference Model Driven Approach*, International Conference on Data and Knowledge Systems for Manufacturing and Engineering (DKSME '94), Hong Kong.

Singh, V., Weston, R. H., 1994c, *Structured Specification and Construction of Open Manufacturing Control Systems,* Journal of Manufacturing Systems, SME (Society of Manufacturing Engineers), USA.

Singh, V., Weston, R. H., 1996a, *Information models: a precursor to software interoperability,* Journal of Production Planning & Control, UK, Vol. 7, No. 3, pp242-257.

Singh, V., Weston, R. H., 1996b, *Life Cycle Support of Manufacturing Systems based on an Integration of Tools,* Journal of Production Research, UK, Vol. 34, No. 1, pp1-17.

Solberg, J. J., 1989, *Managing Information Complexity in Material Flow Systems*, NATO ASI Series, Springer-Verlag, Vol. F53, pp3-20.

Ssemakula, M. E., 1987, *The role of process planning in the integration of CAD/CAM systems*, Procs. of Fourth European Conference on Automated Manufacturing (AUTOMAN 4), Birmingham, UK.

Struedel, H. J., Desruelle, P., 1992, *Manufacturing in the Nineties*, Van Nostrand, Reinhold, New York.

Taylor, F. W., 1947, *Scientific Management*, Harper and Row.

Taylor, R. W., Frank, R. L., 1976, *CODASYL data base management systems*, ACM Computing Surveys, New York, Vol. 8, No. 1.

Terry, W. R., Matz, T. W., 1989, *An object-oriented programming paradigm for synchronous manufacturing*, Computers Industrial Engineering, Vol. 17, Nos 1-4, pp124-129.

Thompson, G. R., Gomer, T., Chung, C., Barkmeyer, E., Carter, F., Templeton, M., Fox, S., Hartman, B., September 1990, *Heterogeneous Distributed Databases Systems for Production Use*, ACM Computing Surveys, Vol. 22, No. 3, pp237-265.

Timon, F., Jagdev, H. S., Browne, J., 1990, *The Analysis of and the selection Criterion for Production Management Packages*, Advances in production management systems, Procs. Fourth International IFIP Conference TC5/WG 5-7, Espoo, Finland, pp427-438.

Van der Lans, R. F., 1989, *The SQL standard*, Prentice Hall.

Van Donselaar, K., July-September 1992, *The use of MRP and LRP in a stochastic environment*, International Journal Production Planning & Control, Vol. 3, No. 3, pp239-246.

Vollmann, T. E., Berry, W. L., Whybark, D. C., 1988, *Manufacturing Planning and Control Systems*, Dow Jones Irwin, Homewood, IL, USA.

Waterlow, J. G., Monniott, J. P., 1986, *A study of the state of the Art in Computer-Aided Production Management in UK industry*, ACME Report, UK.

Weber, D. M., Moodie, C. L., 1989, *Distributed intelligent information systems for automated integrated manufacturing systems*, Advanced Information Technologies for Industrial Material Flow systems, Springer-Verlag.

Weinberg, J. C., 1989, *Linking the CIM Plan with Operations Strategy*, Procs. Conference Autofact' 89, Detroit, MI, USA.

Welz, F., March 1993, *Software Interoperability within Manufacturing Control Systems*, Dept of Manufacturing Engineering, Loughborough University of Technology), LUT Press, UK.

Weston, R. H., 1993, *Steps Towards Enterprise-Wide Integration: a Definition of Need and First Generation Open Solutions*, International Journal of Production Research, Vol. 31, No. 9, pp2235-2254.

Weston, R. H., Gascoigne, J. D., Rui, A., Hodgson, A., Sumpter, C. M., Coutts, I., 1988, *Steps towards information integration in manufacturing*, International Journal Computer Integrated Manufacturing, Vol. 1, No. 3, p140.

Weston, R. H., Zhang, P., Murgatroyd, I. S., Coutts, I. A., Hodgson, A., September 1991, *Soft Integrated Assembly Systems*, Procs. Fourth World Conference on Robotics Research, Pittsburg, PA, USA, pp410-419.

Weymont, N., P., Honeyager, J. S., 1987, *Developing a CIM Architecture*, Procs. of the Digital Equipment Computer Users Society, USA.

White, C. J., Winter 1993/1992, *Interoperability: The Impact of New Standards*, INFODB, Vol. 7, Part 1, pp21-30.

Wight, O., 1984, *Manufacturing Resource Planning: MRP II*, Essex Junction, Oliver Wight Publications Ltd., UK.

Wilkinson, G., G., Winterflood, A. R., 1987, *Fundamentals of Information Technology*, John Wiley and Sons, pp207-219.

Williams, J., Rogers, P., 1991, *Manufacturing cells: control, programming and integration*, Butterworth-Heinemann.

Wood, P. J., Johnson, P. N., 1989, *A review of the use of SSADM and IDEF at the University of Warwick*, Procs. of SAMT' 89 Conference, Sunderland, UK.

Wyatt, T., Al-Maliki, l., 1990, *Methods in manufacturing systems engineering - the background*, Integrated Manufacturing Systems, Vol. 1, No. 2, pp91-93.

Yeomans, R. W., Choudry, A., 1986, *Design Rules for CIM*, North Holland.

Zäpfel, G., Missbauer, H., 1993, *New Concepts for production planning and control*, European Journal of Operational Research, Vol. 67, pp297-320.

Zhang, H. -C., Alting, L., *An Exploration of Simultaneous Engineering for Manufacturing Enterprises*, International Journal of Advanced Manufacturing Technology, Vol. 7, No. 2, pp101-108.

# Index

Activity-based modelling, 118
Application shell, 106

Bespoke interfaces, 18
Business process re-engineering, 39

CIM-BIOSYS IIS, 66
CIM model, 56
CIM-OSA, 39
Client/server model, 24
Configurator, 103
CORBA, 31

Data access middleware, 30
Database access approach, 85
Database 'driver', 87
Database schema, 82
Distributed computing environment, 23
Distributed transaction processing, 30

ELMS CAPM, 73
Engineering data, 92
Engineering resource management, 102
Enterprise modelling, 34
Entity-relationship modelling, 38
EXPRESS, 115
External schema, 83

Function -information association, 41, 99
Functional interaction, 98, 100

Global schema, 83
GRAI methodology, 39

Hierarchical data model, 22

IDEF methodologies, 38, 112
IDEF$_{0/1X}$ Parser, 119
Import and export filters, 20
Information models, 53, 61
Integrated information systems, 20
Integrated life-cycle support, 111
Integrated manufacturing systems, 10
Integrating infrastructure, 27, 64
Integration mechanisms and tools, 64
Interconnection facilities, 18
Internal schema, 83
Islands of computerisation, 5

JAVA, 33
JAVA Virtual Machine, 33
JIT, 51

Legacy systems, 13
Levels of integration, 9
Life-cycle phases, 37, 111

MANDATE, 36
Manufacturing continuum, 49
Manufacturing information, 48
Manufacturing paradigm shift, 7
MCC CAPM, 60
Message-oriented middleware, 28
Middleware services, 28
Model enactment, 44, 115
MRP II, 50

Network computing architecture, 27
Network data model, 22

ODBC, 30
Object-oriented analysis and design, 39
Object request broker architecture, 30
OpenDOC, 31
Open Software Foundation, 27
OPT, 51

'Pair-wise' integration, 18
Product data management, 93

RDBMS, 82
Reference models, 73
Relational data model, 23
Remote procedure call, 24, 29

Software interoperability, 9, 14, 149
SQL, 85
Standards continuum, 11
System behaviour, 41, 98, 119
Systems design and development, 36
Systems Integration Manager, 93
System Object Model (SOM), 31
System-wide data repository, 78

Taylorism, 5
Triple diagonal concept, 41

Workflow management, 93

'Y-CIM' model, 41
Yourdon, 38